Honda XL/XR 250 & 500 Owners Workshop Manual

T0261065

by Pete Shoemark

Models covered

XL250 S. 248cc. UK May 1978 to Sept 1982, US April 1978 to 1981
XL250 R. 248cc. UK Mar 1982 to Oct 1984, US Feb 1982 to 1983
XR250. 248cc. UK Mar 1979 to 1980, US Sept 1978 to 1980
XR250 R. 248cc. US only Sept 1981 to 1982
XL500 S. 498cc. UK Mar 1979 to Feb 1982, US Feb 1979 to 1981
XL500 R. 498cc. UK Feb 1982 to Oct 1984, US July 1981 to 1982
XR500. 498cc. UK Mar 1979 to 1980, US Jan 1979 to 1980
XR500 R. 498cc. US only August 1981 to 1982

ISBN 978 1 85010 268 7

© Haynes Publishing 2001

ABCDE
FGHIJ

2

Printed in the UK   *(567-6AA6)*

British Library Cataloguing in Publication Data
A catalogue record for this book is available from
the British Library

Library of Congress Control Number 86-80781

Haynes Publishing Group
Sparkford Nr Yeovil
Somerset BA22 7JJ   England

Haynes Publications, Inc
859 Lawrence Drive
Newbury Park
California 91320 USA

Acknowledgements

Our thanks are due to Paul Branson Motorcycles of Yeovil and Mike Haimes of Yeovil, who supplied the machines featured in the photographs throughout this manual. P. R. Taylor and Sons of Calne provided the XL250 R shown on the front cover.

The Avon Rubber Company provided information on tyre fitting. NGK Spark Plugs (UK) Limited furnished information on spark plug condition, and Renold Limited assisted with details on chain care and replacements.

About this manual

The purpose of this manual is to present the owner with a concise and graphic guide which will enable him to tackle any operation from basic routine maintenance to a major overhaul. It has been assumed that any work will be undertaken without the luxury of a well-equipped workshop and a range of manufacturer's service tools.

To this end, the machine featured in the manual was stripped and rebuilt in our own workshop, by a team comprising a mechanic, a photographer and the author. The resulting photographic sequence depicts events as they took place, the hands shown being those of the author and the mechanic.

The use of specialised, and expensive, service tools was avoided unless their use was considered to be essential due to risk of breakage or injury. There is usually some way of improvising a method of removing a stubborn component, provided that a suitable degree of care is exercised.

The author learnt his motorcycle mechanics over a number of years, faced with the same difficulties and using similar facilities to those encountered by most owners. It is hoped that this practical experience can be passed on through the pages of this manual.

Where possible, a well-used example of the machine is chosen for a workshop project, as this highlights any areas which might be particularly prone to giving rise to problems. In this way, any such difficulties are encountered and resolved before the text is written, and the techniques used to deal with them can be incorporated in the relevant Section. Armed with a working knowledge of the machine, the author undertakes a considerable amount of research in order that the maximum amount of data can be included in this manual.

Each Chapter is divided into numbered Sections. Within these Sections are numbered paragraphs. Cross reference throughout the manual is quite straightforward and logical. When reference is made 'See Section 6.10' it means Section 6, paragraph 10 in the same Chapter. If another Chapter were intended the reference would read, for example, 'See Chapter 2, Section 6.10'. All the photographs are captioned with a section/paragraph number to which they refer and are relevant to the Chapter text adjacent.

Figures (usually line illustrations) appear in a logical but numerical order, within a given Chapter. Fig. 1.1 therefore refers to the first figure in Chapter 1.

Left-hand and right-hand descriptions of the machine and their components refer to the left and right of a given machine when the rider is seated normally.

Motorcycle manufacturers continually make changes to specifications and recommendations, and these, when notified, are incorporated into our manuals at the earliest opportunity.

We take great pride in the accuracy of information given in this manual, but motorcycle manufacturers make alterations and design changes during the production run of a particular motorcycle of which they do not inform us. No liability can be accepted by the authors or publishers for loss, damage or injury caused by any errors in, or omissions from, the information given.

Contents

Honda XL 250S

Right-hand view of the 1980 Honda XL250 S

Right-hand view of the 1980 Honda XR500

Engine and gearbox unit of the Honda XL250 S

Introduction to the Honda XL250 S, XL500 S, XR250 and XR500

The models covered by this manual represent Honda's latest offerings in the mid-to-large capacity trail and enduro market. The four-stroke single-cylinder machine all but vanished during the 1970s, when light two-stroke machines dominated almost every off-road class, but has now been re-discovered and is rapidly gaining popularity.

The current XL and XR models were introduced between May 1978 and March 1979 and replaced the previous series of models. The machines were substantially revised, the 250 and 500cc machines featuring redesigned engine units with a cleverly contrived balancer system. This arrangement was designed to counter the out-of-balance forces inherent in a single-cylinder engine, these being considered undesirable. Apart from obvious differences in capacity and state of tune, the engine units are of similar design and construction.

The frame is similar on all models, having a top spine section fabricated from pressed and welded sheet steel sections. The engine unit is employed as a stressed member of the frame assembly, and thus the weight and added bulk of an engine cradle can be dispensed with. The remainder of the frame is of conventional welded steel tube construction.

Front suspension is provided by conventional oil-damped telescopic forks, these being of the fashionable 'leading axle' type in which the wheel spindle axis is forward of the lower leg to give greater wheel travel. Rear suspension is currently of conventional twin gas/oil suspension unit and swinging arm construction. All models are equipped with Honda's now famous 23 in front wheel, rather than the standard off-road size of 21in.

The XL and XR versions share many similarities but are intended for very different applications. The XLs are by definition a compromise. Like all trail bikes they are fully equipped for road use but have the capacity to cope with gentle off-road forays. Unlike many trail bikes they succeed in fulfilling these rather conflicting requirements. The four-stroke engine unit produces a wide spread of power when compared with two-strokes of similar capacity, and the compromise between road and off-road gearing is less obvious in either element. The XR machines are off-road machines with few concessions to road use. The limited electrical system suffices to make them road-legal for registration purposes, but little else. It should be noted that the XR as imported into the UK has a number of features which make it illegal to use on a public road, and these are discussed in Chapter 6. Machines registered in the UK will have been modified by the dealer or owner to comply with UK legal requirements, and as no official conversion exists, these modifications cannot be covered in this manual.

Model dimensions and weights

	XL250 S	XR250	XL500 S	XR500
Overall length	2245 mm (88.4 in)	2210 mm (87.0 in)	2175 mm (85.6 in)	2210 mm (87.0 in)
Overall width	870 mm (34.3 in)	875 mm (34.4 in)	890 mm (35.0 in)	875 mm (34.4 in)
Overall height	1185 mm (46.7 in)	1210 mm (47.6 in)	1185 mm (46.7 in)	1215 mm (47.8 in)
Wheelbase	1390 mm (54.7 in)	1400 mm (55.1 in)	1405 mm (55.3 in)	1400 mm (55.1 in)
Seat height	845 mm (33.3 in)	880 mm (34.6 in)	860 mm (33.9 in)	880 mm (34.6 in)
Footrest height	335 mm (13.2 in)	355 mm (14.0 in)	335 mm (13.2 in)	355 mm (14.0 in)
Castor	61° 30'	61° 30'	61° 30'	61° 30'
Trail	138 mm (5.4 in)	138 mm (5.4 in)	138 mm (5.4 in)	138 mm (5.4 in)
Dry weight	123 kg (271 lb)	115 kg (253.5 lb)	132 kg (291.0 lb)	123 kg (271.2 lb)
Ground clearance	260 mm (10.2 in)	280 mm (11.0 in)	260 mm (10.2 in)	280 mm (11.0 in)

Note: slight variations in some of these figures may be found, depending on the country or state of original delivery and year of manufacture.

Ordering spare parts

When ordering spare parts for any Honda model it is advisable to deal direct with an official Honda agent, who should be able to supply most items ex-stock. Parts cannot be obtained from Honda (UK) Limited direct; all orders must be routed via an approved agent, even if the parts required are not held in stock.

Always quote the engine and frame numbers in full, particularly if parts are required for any of the earlier models.

The frame number is located on the right-hand side of the steering head and the engine number is stamped on the upper crankcase, immediately to the rear of the cylinder. Use only parts of genuine Honda manufacture. Pattern parts are avail-

able, some of which originate from Japan, but in many instances they may have an adverse effect on performance and/or reliability. Furthermore the fitting of non-standard parts may invalidate the warranty. Honda do not operate a 'service exchange' scheme.

Some of the more expendable parts such as sparking plugs, bulbs, tyres, oils and greases etc., can be obtained from accessory shops and motor factors, who have convenient opening hours and can often be found not far from home. It is also possible to obtain parts on a Mail Order basis from a number of specialists who advertise regularly in the motor cycle magazines.

Location of frame number

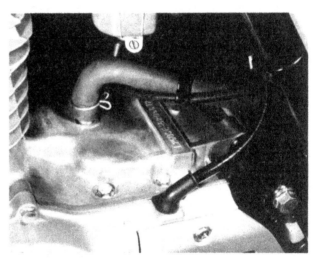

Location of engine number

Location of identification plate

Safety first!

Professional motor mechanics are trained in safe working procedures. However enthusiastic you may be about getting on with the job in hand, do take the time to ensure that your safety is not put at risk. A moment's lack of attention can result in an accident, as can failure to observe certain elementary precautions.

There will always be new ways of having accidents, and the following points do not pretend to be a comprehensive list of all dangers; they are intended rather to make you aware of the risks and to encourage a safety-conscious approach to all work you carry out on your vehicle.

Essential DOs and DON'Ts

DON'T start the engine without first ascertaining that the transmission is in neutral.

DON'T suddenly remove the filler cap from a hot cooling system – cover it with a cloth and release the pressure gradually first, or you may get scalded by escaping coolant.

DON'T attempt to drain oil until you are sure it has cooled sufficiently to avoid scalding you.

DON'T grasp any part of the engine, exhaust or silencer without first ascertaining that it is sufficiently cool to avoid burning you.

DON'T allow brake fluid or antifreeze to contact the machine's paintwork or plastic components.

DON'T syphon toxic liquids such as fuel, brake fluid or antifreeze by mouth, or allow them to remain on your skin.

DON'T inhale dust – it may be injurious to health (see *Asbestos* heading).

DON'T allow any spilt oil or grease to remain on the floor – wipe it up straight away, before someone slips on it.

DON'T use ill-fitting spanners or other tools which may slip and cause injury.

DON'T attempt to lift a heavy component which may be beyond your capability – get assistance.

DON'T rush to finish a job, or take unverified short cuts.

DON'T allow children or animals in or around an unattended vehicle.

DON'T inflate a tyre to a pressure above the recommended maximum. Apart from overstressing the carcase and wheel rim, in extreme cases the tyre may blow off forcibly.

DO ensure that the machine is supported securely at all times. This is especially important when the machine is blocked up to aid wheel or fork removal.

DO take care when attempting to slacken a stubborn nut or bolt. It is generally better to pull on a spanner, rather than push, so that if slippage occurs you fall away from the machine rather than on to it.

DO wear eye protection when using power tools such as drill, sander, bench grinder etc.

DO use a barrier cream on your hands prior to undertaking dirty jobs – it will protect your skin from infection as well as making the dirt easier to remove afterwards; but make sure your hands aren't left slippery. Note that long-term contact with used engine oil can be a health hazard.

DO keep loose clothing (cuffs, tie etc) and long hair well out of the way of moving mechanical parts.

DO remove rings, wristwatch etc, before working on the vehicle – especially the electrical system.

DO keep your work area tidy – it is only too easy to fall over articles left lying around.

DO exercise caution when compressing springs for removal or installation. Ensure that the tension is applied and released in a controlled manner, using suitable tools which preclude the possibility of the spring escaping violently.

DO ensure that any lifting tackle used has a safe working load rating adequate for the job.

DO get someone to check periodically that all is well, when working alone on the vehicle.

DO carry out work in a logical sequence and check that everything is correctly assembled and tightened afterwards.

DO remember that your vehicle's safety affects that of yourself and others. If in doubt on any point, get specialist advice.

IF, in spite of following these precautions, you are unfortunate enough to injure yourself, seek medical attention as soon as possible.

Asbestos

Certain friction, insulating, sealing, and other products – such as brake linings, clutch linings, gaskets, etc – contain asbestos. *Extreme care must be taken to avoid inhalation of dust from such products since it is hazardous to health.* If in doubt, assume that they *do* contain asbestos.

Fire

Remember at all times that petrol (gasoline) is highly flammable. Never smoke, or have any kind of naked flame around, when working on the vehicle. But the risk does not end there – a spark caused by an electrical short-circuit, by two metal surfaces contacting each other, by careless use of tools, or even by static electricity built up in your body under certain conditions, can ignite petrol vapour, which in a confined space is highly explosive.

Always disconnect the battery earth (ground) terminal before working on any part of the fuel or electrical system, and never risk spilling fuel on to a hot engine or exhaust.

It is recommended that a fire extinguisher of a type suitable for fuel and electrical fires is kept handy in the garage or workplace at all times. Never try to extinguish a fuel or electrical fire with water.

Note: *Any reference to a 'torch' appearing in this manual should always be taken to mean a hand-held battery-operated electric lamp or flashlight. It does **not** mean a welding/gas torch or blowlamp.*

Fumes

Certain fumes are highly toxic and can quickly cause unconsciousness and even death if inhaled to any extent. Petrol (gasoline) vapour comes into this category, as do the vapours from certain solvents such as trichloroethylene. Any draining or pouring of such volatile fluids should be done in a well ventilated area.

When using cleaning fluids and solvents, read the instructions carefully. Never use materials from unmarked containers – they may give off poisonous vapours.

Never run the engine of a motor vehicle in an enclosed space such as a garage. Exhaust fumes contain carbon monoxide which is extremely poisonous; if you need to run the engine, always do so in the open air or at least have the rear of the vehicle outside the workplace.

The battery

Never cause a spark, or allow a naked light, near the vehicle's battery. It will normally be giving off a certain amount of hydrogen gas, which is highly explosive.

Always disconnect the battery earth (ground) terminal before working on the fuel or electrical systems.

If possible, loosen the filler plugs or cover when charging the battery from an external source. Do not charge at an excessive rate or the battery may burst.

Take care when topping up and when carrying the battery. The acid electrolyte, even when diluted, is very corrosive and should not be allowed to contact the eyes or skin.

If you ever need to prepare electrolyte yourself, always add the acid slowly to the water, and never the other way round. Protect against splashes by wearing rubber gloves and goggles.

Mains electricity and electrical equipment

When using an electric power tool, inspection light etc, always ensure that the appliance is correctly connected to its plug and that, where necessary, it is properly earthed (grounded). Do not use such appliances in damp conditions and, again, beware of creating a spark or applying excessive heat in the vicinity of fuel or fuel vapour. Also ensure that the appliances meet the relevant national safety standards.

Ignition HT voltage

A severe electric shock can result from touching certain parts of the ignition system, such as the HT leads, when the engine is running or being cranked, particularly if components are damp or the insulation is defective. Where an electronic ignition system is fitted, the HT voltage is much higher and could prove fatal.

Routine maintenance

Periodic routine maintenance is a continuous process that commences immediately the machine is used. It must be carried out at specified mileage recordings or on a calendar date basis if the machine is not used regularly, whichever falls sooner. Maintenance should be regarded as an insurance policy, to help keep the machine in peak condition and to ensure long, trouble-free service. It has the additional benefit of giving early warning of any faults that may develop and will act as a regular safety check, to the obvious advantage of both rider and machine alike.

The nature of the various models covered in this manual calls for a rather more flexible approach to routine maintenance than would be applicable to road machines. The XL models may be used primarily for road work, with occasional off-road forays, whilst the XR versions will invariably spend most of their working lives in particularly inclement conditions. It follows that the service intervals quoted can serve only as a basic guide, and should be adjusted to suit local conditions and climate.

Owners of XL machines should refer to the first of the maintenance schedules, which gives a summary of the operations involved with a mileage and calendar heading. Each entry includes a reference number which precedes the detailed description of the operation in the sections which follow. The second schedule relates to general maintenance for the XR enduro machines, and is keyed to the text in a similar manner. A further table is provided listing those items which should be attended to prior to every competitive event. It should be remembered that the interval between the various maintenance tasks serves only as a guide. As the machine gets older or is used under particularly adverse conditions, it would be advisable to reduce the period between each check.

For ease of reference each service operation is described in detail under the relevant heading. However, if further general information is required, it can be found within the manual under the pertinent section heading in the relevant Chapter.

In order that the routine maintenance tasks are carried out with as much ease as possible, it is essential that a good selection of general workshop tools is available.

Included in the kit must be a range of metric ring or combination spanners, a selection of crosshead screwdrivers and at least one pair of circlip pliers.

Additionally, owing to the extreme tightness of most casing screws on Japanese machines, an impact screwdriver, together with a choice of large and small crosshead screw bits, is absolutely indispensable. This is particularly so if the engine has not been dismantled since leaving the factory.

Maintenance schedule – XL models

Operation	Interval/Section reference				
	Weekly 125 miles 200 km	Monthly 600 miles 800 km	3 monthly 1800 miles 2400 km	6 monthly 3600 miles 4800 km	Yearly 7200 miles 9600 km
Check engine/transmission oil level	1				
Control adjustments and cables	2				
Tyre condition and pressures	3				
Lights and instruments	4				
Lubricate controls cables and pivots		5			
Clean lubricate and adjust chain		6			
Check and adjust valve clearances		7			
Adjust decompressor cable		8			
Check and adjust cam chain		9			
Check battery condition		10			
Check and adjust brakes		11			
Check and adjust clutch		12			
Check steering for play		13			
Check suspension operation		14			
Check idle speed and throttle setting		15			
Clean air filter element		16			
Change engine/transmission oil			17		
Clean oil filter screen				18	
Clean and adjust sparking plug				19	
Check and adjust balancer					20
Renew sparking plug					21

Maintenance schedule – XR models

The following operations should be carried out after the initial 200 miles (350 km) and thereafter at 1000 mile (1600 km) intervals unless marked otherwise.

Operation	Section	Operation	Section
Check engine/transmission oil level	1	Check and adjust clutch	12
Control adjustments and cables	2	Check steering for play	13
Tyre condition and pressures	3	Check suspension operation	14
Lights and instruments	4	Check idle speed and throttle setting	15
Lubricate controls cables and pivots	5	Clean air filter element	16*
Clean lubricate and adjust chain	6*	Change engine/transmission oil	17**
Check and adjust valve clearances	7	Clean oi filter screen	18
Adjust decompressor cable	8	Clean and adjust sparking plug	19
Check and adjust cam chain	9	Check and adjust balancer	20
Check battery condition	10	Renew sparking plug	21
Check and adjust brakes	11		

*   Clean at 500 mile (800 km) intervals, or more often in dusty conditions
** Change at 1800 mile (3000 km) intervals

Competition maintenance checks – XR models

In addition to the normal routine maintenance the following check list should be used prior to every competitive event.

Item	Check for	Action (as required)	Section
Engine oil	Level, contamination	Top up/change	1, 17
Fuel system	Leaks or damage	Repair renew	–
Carburettor	Operation	Adjust or clean	15
Air filter	Dirt or damage	Clean or replace	16
Sparking plug	Damage or contamination	Clean, adjust or renew	19, 21
Valve clearances	Correct clearance	Adjust	7
Balancer chain	Correct tension	Adjust	20
Decompressor	Correct tension	Adjust	8
Cam chain	Correct tension	Adjust	9
Drive chain	Contamination and tension	Clean and adjust	6
Sprockets	Wear and damage	Check and renew	–
Brakes	Wear and adjustment	Adjust/renew	11
Lights	Condition	Check and adjust	4
Suspension	Free travel	Check	14
Steering	Play, free movement	Check	13
Clutch	Condition, adjustment	Check	12
Swinging arm	Free play	Lubricate/overhaul	–
Controls, cables	Adjustment, condition	Adjust, lubricate	2
Nuts and bolts	Check tightness	Tighten	–

1 Checking the engine/gearbox oil level

Unscrew the filler plug which is situated to the rear of the right-hand crankcase half; it will be noted that the plug incorporates a dipstick which should be wiped off using a clean, lint-free rag. Place the plug back in position, but do not screw it home, allow it to rest in position on the edge of the orifice.

Remove the plug and note the level of the oil on the dipstick, which should be between the two level marks. If necessary, top up using SAE 10W/40 engine oil. Note that if the machine has been ridden recently, it should be allowed to stand for a few minutes to allow the oil clinging to the internal surfaces to drain down into the sump. Also, the machine should be vertical during the check so that an accurate oil level reading is obtained.

2 Checking the controls and cables

Carefully examine each of the control levers, pedals and cables, looking for signs of wear or damage which might indicate imminent failure. Whilst this is of obvious importance where the machine is used in competition, it applies equally to the XL models where a failure on or off the road can be inconvenient or even dangerous. Remove any accumulated dirt from around the pivots and cable ends, and check that the protective cable boots are intact and in position. Check that the front brake lever and rear brake pedal are correctly adjusted and the kickstart and gearchange levers are secure. Lubricate the

exposed portions of the cables and all pivots with a multi-purpose aerosol lubricant such as WD40 or similar.

3 Tyre pressures and condition

Check the tyre pressures with a gauge that is known to be accurate. It is worthwhile purchasing a pocket gauge for this purpose, as the gauges on garage forecourts cannot always be relied upon as being accurate. The readings should not be taken after the machine has been used, as the tyres will have become warm. This will have caused the pressure to increase, giving a false reading.

Tyre pressures

XL models	Solo	With pillion
Front	21 psi (1.5 kg cm^2)	21 psi (1.5 kg cm^2)
Rear	21 psi (1.5 kg cm^2)	25 psi (1.75 kg cm^2)
XR models		
Front	14 psi (1.0 kg cm^2)	–
Rear	17 psi (1.2 kg cm^2)	–

Give the tyres a close visual check, looking for signs of damage to the tread and sidewalls. Remove any stones caught between the tread blocks. The wheel rims and spokes should also be given a quick check, especially after off-road use where impact damage may have been sustained.

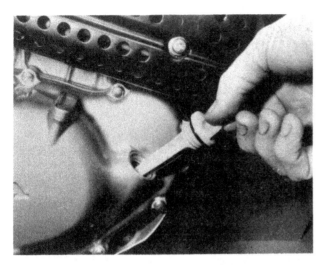

Oil filler plug incorporates dipstick

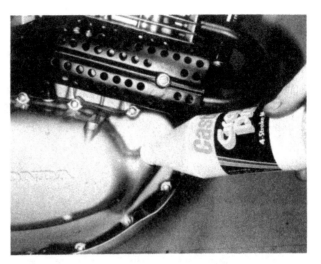

Top up engine/transmission oil to prescribed level

4 Lights and instruments

With the engine running check that the headlamp and tail lamp operate correctly (all models) and that the parking lamp, indicators and horn function (XL models only). When riding, check that the speedometer and odometer are operating. These checks can be carried out each time the machine is ridden, forming a routine check rather than a specific maintenance operation.

5 Lubricating controls, cables and pivots

Carry out the general checks and cleaning as described in Section 2, paying particular attention to the ends of the control cables. Any signs of kinking or fraying will indicate that renewal is required. To obtain maximum life and reliability from the cables they should be thoroughly lubricated. To do the job properly and quickly use one of the hydraulic cable oilers available from most motorcycle shops. Free one end of the cable and assemble the cable oiler as described by the manufacturer's instructions. Operate the oiler until oil emerges from the lower end, indicating that the cable is lubricated throughout its length. This process will expel any dirt or moisture and will prevent its subsequent ingress.

If a cable oiler is not available, an alternative is to remove the cables from the machine. Hang the cable upright and make up a small funnel arrangement using plasticine or by taping a plastic bag around the upper end. Fill the funnel with oil and leave it overnight to drain through. Note that where nylon-lined cables are fitted, they should be used dry or lubricated with a silicone-based lubricant suitable for this application. On no account use ordinary engine oil because this will cause the liner to swell, pinching the cable.

Check all pivots and control levers, cleaning and lubricating them to prevent wear or corrosion. Where necessary, dismantle and clean any moving part which may have become stiff in operation.

6 Final drive chain — cleaning, lubrication and adjustment

In order that final drive chain life can be extended as much as possible, regular lubrication and adjustment is essential. This is particularly so when the chain is not enclosed or is fitted to a machine transmitting high power to the rear wheel. The chain may be lubricated whilst it is in place on the machine by the application of one of the proprietary chain greases contained in an aerosol can. Ordinary grease oil can be used, though owing to the speed with which it is flung off the rotating chain, its effective life is limited.

The most satisfactory method of chain lubrication can be made when the chain has been removed from the machine, because this allows it to be cleaned thoroughly prior to lubrication. Unfortunately, the chain fitted as original equipment to the XL models is of the endless variety, which makes it necessary to remove the rear wheel and swinging arm prior to chain removal. Failing this, support the machine on blocks so that the rear wheel is raised clear of the ground. Clean the chain using a paraffin/petrol mix, making sure that all accumulated mud and road dirt is removed from the links and rollers. When the chain is clean and dry use a chain spray to lubricate and protect the chain. Note that the above procedure, if performed with sufficient frequency, will prolong the life of chains considerably, especially where the machine is used off road. On XR models the chain may be removed after separating it at the spring link. Clean the chain in a petrol/paraffin mix and allowing it to dry before immersing it in one of the special chain greases such as Linklyfe or Chainguard. After lubrication install the chain, ensuring that the spring link is fitted correctly with the closed end of the link spring facing in the direction of normal travel.

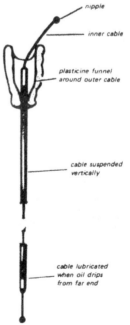

nipple

inner cable

plasticine funnel around outer cable

cable suspended vertically

cable lubricated when oil drips from far end

Control cable oiling

Aerosol chain lubricant is used after cleaning

XL models

Check the slack in the final drive chain. The correct up and down movement, as measured at the mid-point of the chain lower run, should be 30 – 40 mm (1¼ – 1⅝ in) with the machine on the side stand. Remove the split pin from the wheel spindle and slacken the wheel nut a few turns. Loosen the locknuts on the two chain adjuster bolts. Rotation of the adjuster bolts in a clockwise direction will tighten the chain. Tighten each bolt a similar number of turns so that wheel alignment is maintained. This can be verified by checking that the mark on the outer face of each chain adjuster is aligned with the same aligning mark on each fork end. With the adjustment correct, tighten the wheel nut and fit a new split pin. Finally, retighten the adjuster bolt locknuts.

XR models

The procedure for the XR models is similar to that described above, but play should be measured by pushing the lower run of the chain towards the swinging arm. The normal clearance is 30 – 40 mm (1¼ – 1⅝ in).

With all models, check the condition of the chain tensioner or guide blocks, lubricating or renewing these as required.

7 Checking and adjusting the valve clearances

The accurate setting of valve clearances is essential if the engine is to function properly. If the clearance becomes too great, the valves will not open fully. This restricts the amount of fuel/air mixture entering the cylinder which in turn lessens the power produced on each firing stroke. The result is a noisy and inefficient engine. Conversely, too small a clearance will mean that the valves do not close fully, leading to a marked fall-off in performance. More significantly, the escaping hot gases will quickly destroy the valve faces.

The valve clearances are set with the engine cold, which is best interpreted as after the machine has stood overnight, ensuring that it has cooled fully since it was last run. If the prescribed clearance is set at this engine temperature (below 95°F, 35°C) it will ensure that the valves open fully and close fully when the engine is at normal running temperature.

Start by removing the seat, which is secured by a bolt on each side. Lift the seat clear, then remove the single fixing bolt at the rear of the fuel tank. Check that the fuel tap is off and pull off the fuel feed pipe. The tank can now be pulled rearwards and removed.

The engine must be at the top dead centre (TDC) position on the compression stroke, and to this end the alternator rotor has a T (Timing) mark which should be aligned with the fixed index mark on the cover. Remove the two inspection caps from the left-hand outer cover and release the valve inspection covers. Remove the sparking plug so that the crankshaft can be rotated easily. Using a socket passed through the central inspection hole in the left-hand outer cover, turn the engine anti-clockwise until the T mark appears in the upper hole with the adjacent timing mark aligned with the index mark.

Check that the engine is at TDC compression by ensuring that none of the valves are open. If this is not the case, turn the engine through another 360° and re-align the T mark. At this position there should be detectable clearance between the valve stems and rockers. Measure this clearance using successive feeler gauges until the gap is known. The gauge should be a light sliding fit between the valve and rocker. Note the clearance of each valve, and if necessary adjust to the correct clearance which is shown below. Adjustment is carried out after the locknut has been slackened. Turn the square-headed adjuster until the required clearance is obtained, then hold the adjuster position whilst the locknut is secured. If required, a Honda dealer can supply a special tool to hold the small square adjuster head, although most owners will be able to improvise with a small open-ended or adjustable spanner. After adjustment, turn the crankshaft through several revolutions and re-check the setting before refitting the covers, caps, tank and seat.

Inspection covers provide access to valve adjusters

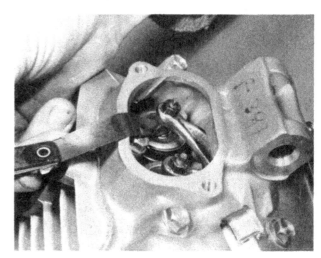

Check valve clearances using feeler gauges

Valve clearances (with cold engine)
XL250 S, XL500 S and XR500
    Inlet   0.05 mm (0.002 in)
    Exhaust 0.10 mm (0.004 in)
XR250
    Inlet   0.08 mm (0.003 in)
    Exhaust 0.10 mm (0.004 in)

Note that the decompressor cable adjustment must be checked whenever the valve clearances are checked or adjusted. The procedure is described in Section 8.

8 Adjusting the decompressor cable

The XL and XR models covered in this manual are equipped with an automatic decompressor, or valve lifter, as an aid to starting. The decompressor is controlled by a cam arrangement on the kickstart shaft and operates via an adjustable cable. The decompressor lever operates on the exhaust valves, holding them open to allow easy cranking of the engine. If the valve clearances are altered for any reason it is necessary to check that the decompressor lever still has the necessary amount of free play. Incorrect adjustment could result in the system not working correctly. If the adjustment is over-tight, burnt exhaust valves could result.

With the alternator rotor T mark aligned as described in Section 7 and the engine on the compressor stroke, check the clearance in the decompressor mechanism, measured at the cylinder head lever. When correctly adjusted there should be 1 – 3 mm (0.04 – 0.12 in) free play at the lever end. If necessary, slacken the adjuster locknut and turn the adjuster to obtain the correct clearance. Do not omit to secure the locknut.

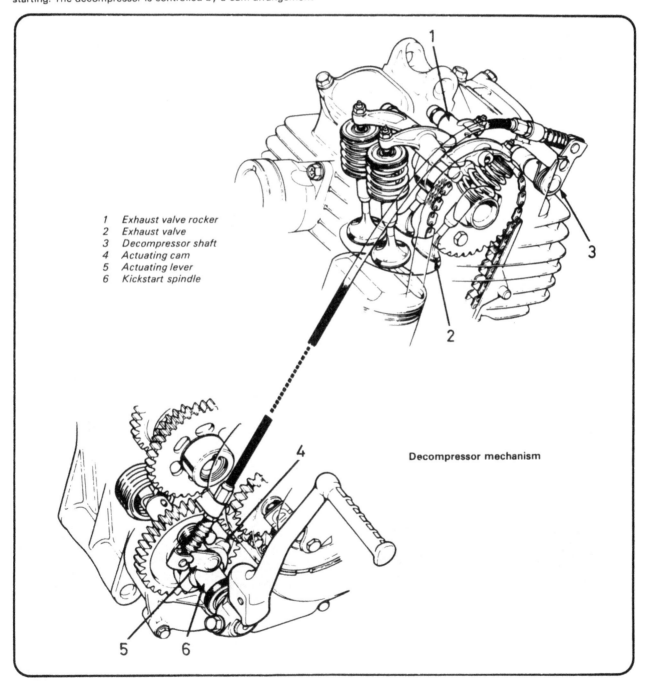

1   Exhaust valve rocker
2   Exhaust valve
3   Decompressor shaft
4   Actuating cam
5   Actuating lever
6   Kickstart spindle

Decompressor mechanism

Decompressor cable adjuster (arrowed)

9 Adjusting the cam chain tension

Over a period of time the cam chain will gradually stretch and wear, allowing excessive free play, and thus noise, to develop. To compensate for this a spring loaded tensioner mechanism is fitted and is used to apply the appropriate pressure to the tensioner blade. The cam chain tension should be reset at the specified intervals or whenever it appears to be excessively nosiy. If cam chain tension adjustment fails to effect a cure it may be because the chain has worn to the point where renewal is necessary. The replacement procedure is covered in Chapter 1.

XL250 S and XR250 models

Start the engine and allow it to idle (1200 ± 100 rpm). Slacken the cam chain tensioner locknut which will be found projecting from the centre of the rear face of the cylinder barrel, immediately below the carburettor. Allow the engine to continue running whilst the tensioner automatically adjusts to the correct position under spring pressure. Secure the locknut to complete the job.

XL500 S and XR500 models

Start the engine and allow it to idle (1200 ± 100 rpm). Identify the two tensioner fasteners, namely a bolt which passes horizontally into the rear face of the cylinder head and a domed

nut located immediately below this. Slacken both by 1½ - 2 turns, noting that further slackening can allow the tensioner to become displaced which in turn will cause extensive engine damage. Allow the tensioner assembly to assume the correct position under spring pressure, then tighten the adjuster bolt and nut.

10 Checking the battery condition

The XL250 S and XL500 S models are equipped with a small 6 volt 4Ah battery which supplies all but the headlamp circuit, this being supplied direct from the alternator. The main function of the battery is to provide a stable supply to the indicator circuit. The battery is housed in a small plastic case which is attached to the rear right-hand side of the frame.

The translucent plastic case of the battery permits the upper and lower levels of the electrolyte to be observed when the left-hand side cover is removed. Maintenance is normally limited to keeping the electrolyte level between the prescribed upper and lower limits and by making sure that the vent pipe is not blocked.

Unless acid is spilt, as may occur if the machine falls over, the electrolyte should always be topped up with distilled water, to restore the correct level. If acid is spilt on any part of the machine, it should be neutralised with an alkali such as washing soda and washed away with plenty of water, otherwise serious corrosion will occur. Top up with sulphuric acid of the correct specific gravity (1.260 - 1.280) only when spillage has occurred. Check that the vent pipe is well clear of the frame tubes or any other of the cycle parts for obvious reasons.

11 Check and adjust the brakes

Check that the brakes operate smoothly and effectively and that they are adjusted correctly. A small amount of free play at the lever or pedal is necessary to ensure that the brake shoes clear the drum face when released, but the exact amount of travel is best left to the owner, who will have a preference for more or less lever or pedal travel. Needless to say, excessive travel which might result in full braking effort not being obtained should be avoided. Note that the rear brake pedal height can be adjusted by means of a locknut and screw at the rear of the pedal pivot.

Check that the linings are within limits by applying each brake fully and noting the position of the wear limit pointer in relation to the index mark on the brake plate. If the two coincide it can be assumed that the brake shoes are in need of renewal. If the shoes are within limits but brake efficiency is impaired, refer to Chapter 5 for details of brake overhaul. Check that the linings are free from oil or water contamination and that the friction surface has not become glazed.

Translucent battery case shows electrolyte level

Rear brake cable incorporates adjuster assembly

12 Check clutch operation and adjustment

Clutch adjustment will be necessary to compensate for wear in the clutch plates and should be carried out at the prescribed interval, or whenever excessive lever travel is evident. The manufacturer recommends that the clutch is adjusted to give 15 – 25 mm ($\frac{5}{8}$ – 1 in) free play at the lever end. If significant adjustment is required it is best to make it at the lower adjuster, the handlebar lever adjuster can then be used for fine adjustment.

If the clutch becomes stiff or jerky in operation, check that the cable is clean, well lubricated and undamaged. Note that a stiff cable can add greatly to the lever pressure required to operate the clutch. Further details on clutch dismantling and overhaul will be found in Chapter 1.

13 Checking the steering head bearings

Wear or play in the steering head bearings will cause imprecise handling and can be dangerous if allowed to develop unchecked. Test for play by pushing and pulling on the handlebars whilst holding the front brake on. Any wear in the head races will be apparent as movement between the head lug and the upper and lower fork yokes.

Before carrying out adjustment, place a wooden crate or similar beneath the skid plate so that the front wheel is raised clear of the ground. Check that the handlebars will turn smoothly and freely from lock to lock. If the steering feels notchy or jerky in operation it may be due to worn or damaged bearings. Should this be suspected it will be necessary to overhaul the steering head bearings as described in Chapter 4.

To adjust the steering head bearings, slacken the large steering stem nut at the centre of the fork top yoke, then use a C-spanner to tighten the slotted adjuster nut immediately below the top yoke. As a guide to adjustment, tighten the slotted nut until a firm resistance is felt, then back it off by $\frac{1}{8}$ turn. The object is to remove all discernible play without applying any appreciable preload. It should be noted that it is possible to apply a loading of several tons on the small steering head bearings without this being obvious when turning the handlebars. This will cause an accelerated rate of wear, and thus must be avoided. Over-tight head races will produce a rolling effect on the machine at very low speeds.

14 Checking the front and rear suspension

Ensure that the front forks operate smoothly and progressively by pumping them up and down whilst the front brake is held on. Any faults revealed by this check should be investigated further, as any deterioration in the handling of the machine can have serious consequences if left unremedied. Check the condition of the fork stanchions. On the XL models the fork stanchions are left exposed as a concession to fashion, and are thus very vulnerable to stone chips when the machine is used off road. There is also a tendency for water to collect inside the short dust seals, and this can promote corrosion of the stanchion or may be drawn past the oil seal where it will contaminate the damping oil. Displace the seals and check that the area is clean and dry and that the fork oil seals are not leaking. Those problems can be eliminated by fitting fork gaiters of the type fitted standard to the XR versions, and this is strongly recommended especially where the XL models are used off-road.

Raise the rear wheel clear of the ground by placing a wooden crate or similar support beneath the machine. Check that all of the suspension components are securely attached to the frame. Check for free play in the swinging arm by pushing and pulling it horizontally. Assuming that all is well, complete the operation by greasing the swinging arm pivot via the grease nipple provided.

15 Checking throttle operation and idle speed

Check that both of the throttle cables are clean and undamaged, ensuring that they are correctly routed and have not become kinked or trapped at any point. The cable adjustment should be set so that there is 2 – 6 mm (0.08 – 0.24 in) free play measured at the flanged inner edge of the rubber twistgrip. Adjustment is effected by means of an adjuster at the lower end of the cable. Further fine adjustment can be made at the upper adjuster.

The manufacturer recommends that the idle speed is checked and set to the prescribed 1200 ± 100 rpm (XR250 model: 1300 ± 100 rpm). This assumes that a tachometer is available for the purposes of the check, since none of the models covered in this manual have a tachometer as standard equipment. Failing this, set the throttle stop screw to give the slowest reliable idle speed with a warm engine. The throttle stop control terminates in a black plastic knob located just to the left of the carburettor.

16 Cleaning the air filter element

It is vitally important that the air filter element is kept clean and in good condition if the engine is to function properly. If the element becomes choked with dust it follows that the airflow to the engine will be impaired, leading to poor performance and high fuel consumption. Conversely, a damaged filter will allow excessive amounts of unfiltered air to enter the engine, which can result in an increased rate of wear and possibly damage due to the weak nature of the mixture. The intervals shown at the front of this Chapter indicate the maximum time limit between each cleaning operation. Where the machine is used under particularly adverse condition it is advised that cleaning takes place on a much more frequent basis.

The element is retained in the air cleaner casing, access to which is via the left-hand side panel. The latter is retained by three screws. In the case of the XL250 S and XL500 S models, the element is held by a spring steel retainer and can be lifted clear after this has been withdrawn. The XR models employ a slightly different retention method, consisting of a wing nut and a screw clip.

Once the element has been removed, peel the foam section off the metal former. The foam can be cleaned by washing in a non-flammable or high flash point solvent. The use of petrol (gasoline) is not approved by the manufacturer in view of the potential fire risk. Allow the element to dry, then impregnate the foam with SAE 80 or 90 gear oil, removing any excess by squeezing it out. The element can now be reassembled and fitted.

If inspection has revealed any holes or tears, the element must be renewed immediately. On no account be tempted to omit the element in view of the damage that may ensue.

17 Changing the engine/transmission oil

The engine and transmission components share a common supply of lubricating oil contained in the unit's wet sump. Under normal conditions, the oil should be changed at the intervals specified at the front of this Chapter, but this interval can be reduced where the machine is used in particularly adverse conditions or for short journeys only.

Obtain a container of about $\frac{1}{2}$ gallon ($2\frac{1}{2}$ litre) capacity into which the old oil can be drained. Place the container in position and remove the drain plug which is situated just to the front of the gearchange pedal shaft. This operation is best carried out while the engine is still hot, because the oil will drain more quickly. Note that the oil filter screen should be removed and cleaned at this stage. See Section 18.

When the oil has drained completely, clean the orifice, drain plug and the sealing washer, ensuring that the latter is undamaged. Refit and tighten the drain plug, then add engine oil to bring the level to between the upper and lower marks on the dipstick. The machine must be supported upright (not on the side stand) when checking the oil level, and the dipstick rested on the filler hole but not screwed fully home. The recommended oil grades and capacity are as follows.

Engine oil capacity
   1.5 litres (1.59 US quarts/2.6 Imp pints)
Engine oil grades
   Normal temperatures   *SAE 10W/40*
   Above 59°F (15°C)   *SAE 30*
   32° – 59°F (0° – 15°C)   *SAE 20 or 20W*
   Below 32°F (0°C)   *SAE 10W*

18 Cleaning the engine oil filter screen

Oil is drawn into the lubrication system via a gauze filter screen located in the right-hand crankcase half. As this is the only form of oil filtration it is essential that the screen is removed for cleaning on a regular basis. It will be appreciated that it is advantageous to carry out this operation in conjunction with each oil change (17) because it is necessary to drain the engine transmission oil before the screen can be removed.

Drain the engine oil as described in Section 17. Remove the skid plate from the underside of the engine and remove the kickstart lever from its shaft. Slacken off their adjusters, then disconnect the clutch cable and decompressor cable from their levers on the engine right-hand outer casing. Slacken off the rear brake cable adjuster nut, then remove the right-hand footrest/brake pedal assembly together with the cable, noting that the brake switch spring should be disconnected as the assembly is removed.

Remove all bolts from the engine right-hand outer casing and remove the casing, noting that there may be some residual oil in the casing. Peel off the old gasket.

The rectangular gauze screen is located horizontally in a slot formed in the crankcase. Pull the screen out of its slot and clean it by washing it in a high flash-point solvent. When clean insert the screen back into the crankcase. Note that it is a good idea to check the balancer chain at this stage because it shares the same service interval.

Refit the engine right-hand outer casing using a new gasket, noting that the decompressor lever must be lifted upwards as the casing is installed to allow its arm on the inside of the casing to engage the track on the kickstart's decompressor cam. When the casing is in place install the bolts and tighten them securely.

Refit the footrest/brake pedal and switch spring, decompressor cable and clutch cable in a reverse of the removal sequence. Refer to the appropriate sections of Routine Maintenance for cable adjustment and set-up details.

Withdraw oil filter screen for cleaning

Install the skid plate, then install the engine oil drain plug and top up the engine with the correct amount and type of oil as described in Section 17. Check the clutch, rear brake and decompressor for correct operation before riding the machine.

19 Cleaning and resetting the sparking plug

Detach the sparking plug cap, and using the correct spanner, remove the sparking plug. Clean the electrodes using a wire brush followed by a strip of fine emery cloth or paper. Check the plug gap with a feeler gauge, adjusting it if necessary to within the range 0.6 – 0.7 mm (0.024 – 0.028 in). Make adjustments by bending the outer electrode, never the inner (central) electrode.

Before fitting the sparking plug smear the threads with a graphited grease; this will aid subsequent removal.

20 Checking and adjusting the balancer chain

The engine unit is fitted with a pair of balancer weights which counteract the normal out-of-balance forces found in a single-cylinder engine. The rear balancer weight runs on the left-hand end of the gearbox mainshaft, whilst the front weight is mounted on the right-hand end of a separate balancer shaft. The two are connected and timed by a single-row chain driven from the left-hand end of the crankshaft.

To facilitate adjustment to compensate for chain wear, the front balancer shaft runs in the eccentric bore of a tubular holder. The right-hand end of the holder terminates in an adjuster plate which is locked by a single retaining bolt. When adjustment is necessary the holder is rotated in the crankcase, effectively moving the axis of the front balancer shaft forwards or backwards.

Balancer chain adjustment requires the removal of the right-hand outer casing, and it is thus convenient to carry out this operation whilst the oil filter screen is being dealt with. Identify the balancer holder lock bolt, which passes through the adjuster plate's elongated slot. Slacken and remove the bolt, noting that the holder should rotate anti-clockwise under spring pressure. Make sure that the holder rotates freely by pulling it back against spring tension and releasing it. This should result in its springing back until the balancer chain is tensioned.

The lower edge of the adjuster plate is marked by a series of lines. The balancer chain tension is set by moving the adjuster plate back (clockwise) by one graduation from the fully tensioned position. Holding this position, refit and tighten the locking bolt to 16 – 20 lbf ft (2.2 – 2.8 kgf m).

Occasionally it may prove impossible to obtain sufficient adjustment within the range of movement provided by the elongated slot. If this proves to be the case it will be necessary to move the adjuster plate in relation to the holder. Disconnect the adjuster spring. Release the circlip which retains the balance weight, then slide the latter off the end of the balancer shaft. Remove the plain washer which is fitted behind the balancer weight, then release the large circlip which secures the adjuster plate to the holder. It will be seen that the adjuster plate is located by tangs which engage in corresponding slots in the holder. Withdraw the adjuster plate and reposition it one slot further round (clockwise) to bring the range of adjustment within the scope of the plate.

Reassemble the balancer components in the reverse of the dismantling sequence, noting that the balancer timing mark must align with its counterpart on the balancer shaft. The tensioning operation can now be completed as described above.

21 Renew the sparking plug

The manufacturer recommends tht the sparking plug is renewed as a precautionary measure at approximately 7000 mile intervals. Always ensure that a plug of the correct type and heat range is fitted, and that the gap is set to the prescribed 0.6 – 0.7 mm (0.024 – 0.028 in) prior to installation. If the old plug is in a reasonable condition, it can be cleaned and re-gapped and carried as an emergency spare in the toolbox.

Quick glance
maintenance adjustments and capacities

For specifications relating to later models, see Chapter 7

	XL250 S and XL500 S	XR250 and XR500
Engine/transmission oil capacity		
Dry	2.0 lit (4.2/3.6 US/Imp pint)	2.0 lit (4.2/3.6 US/Imp pint)
At oil change	1.5 lit (3.2/2.6 US/Imp pint)	1.5 lit (3.2/2.6 US/Imp pint)
Front forks		
Oil capacity (per leg)	185 – 195 cc	200 – 205 cc
Sparking plug gap	0.6 – 0.7 mm (0.024 – 0.028 in)	0.6 – 0.7 mm (0.024 – 0.028 in)
Tyre pressures (cold)		
Front	21 psi (1.5 kg/cm²)	14 psi (1.0 kg/cm²)
Rear:		
Solo	21 psi (1.5 kg/cm²)	17 psi (1.2 kg/cm²)
With passenger	25 psi (1.75 kg/cm²)	–

Note: pressures shown are for road use (XL models) or off-road use (XR models). Pressures may be varied for off-road work, but should be restored to normal when machines are ridden on the public road.

Valve clearances (cold)	Inlet	Exhaust
XR250	0.08 mm (0.003 in)	0.10 mm (0.004 in)
All others	0.05 mm (0.002 in)	0.10 mm (0.004 in)

Recommended lubricants

Components	Lubricant
Engine/transmission	
General, all-temperature use	SAE 10W/40
Above 15°C (60°F)	SAE 30
–10° to + 15°C (15° – 60°F)	SAE 20 or 20W
Above –10°C (15°F)	SAE 20W/50
Below 0°C (32°F)	SAE 10W
Front forks	Automatic transmission fluid (ATF)
Chain	Aerosol chain lubricant or special chain grease
General lubrication	Light machine oil
Wheel bearings	High melting point grease
Swinging arm	High melting point grease
Component lubrication during engine reassembly (see text)	Molybdenum disulphide grease or engine oil as required

Tools and working facilities

The first priority when undertaking maintenance or repair work of any sort on a motorcycle is to have a clean, dry, well-lit working area. Work carried out in peace and quiet in the well-ordered atmosphere of a good workshop will give more satisfaction and much better results than can usually be achieved in poor working conditions. A good workshop must have a clean flat workbench or a solidly constructed table of convenient working height. The workbench or table should be equipped with a vice which has a jaw opening of at least 4 in (100 mm). A set of jaw covers should be made from soft metal such as aluminium alloy or copper, or from wood. These covers will minimise the marking or damaging of soft or delicate components which may be clamped in the vice. Some clean, dry, storage space will be required for tools, lubricants and dismantled components. It will be necessary during a major overhaul to lay out engine/gearbox components for examination and to keep them where they will remain undisturbed for as long as is necessary. To this end it is recommended that a supply of metal or plastic containers of suitable size is collected. A supply of clean, lint-free, rags for cleaning purposes and some newspapers, other rags, or paper towels for mopping up spillages should also be kept. If working on a hard concrete floor note that both the floor and one's knees can be protected from oil spillages and wear by cutting open a large cardboard box and spreading it flat on the floor under the machine or workbench. This also helps to provide some warmth in winter and to prevent the loss of nuts, washers, and other tiny components which have a tendency to disappear when dropped on anything other than a perfectly clean, flat, surface.

Unfortunately, such working conditions are not always available to the home mechanic. When working in poor conditions it is essential to take extra time and care to ensure that the components being worked on are kept scrupulously clean and to ensure that no components or tools are lost or damaged.

A selection of good tools is a fundamental requirement for anyone contemplating the maintenance and repair of a motor vehicle. For the owner who does not possess any, their purchase will prove a considerable expense, offsetting some of the savings made by doing-it-yourself. However, provided that the tools purchased meet the relevant national safety standards and are of good quality, they will last for many years and prove an extremely worthwhile investment.

To help the average owner to decide which tools are needed to carry out the various tasks detailed in this manual, we have compiled three lists of tools under the following headings: *Maintenance and minor repair, Repair and overhaul,* and *Specialized.* The newcomer to practical mechanics should start off with the simpler jobs around the vehicle. Then, as his confidence and experience grow, he can undertake more difficult tasks, buying extra tools as and when they are needed. In this way, a *Maintenance and minor repair* tool kit can be built-up into a *Repair and overhaul* tool kit over a considerable period of time without any major cash outlays. The experienced home mechanic will have a tool kit good enough for most repair and overhaul procedures and will add tools from the specialized category when he feels the expense is justified by the amount of use these tools will be put to.

It is obviously not possible to cover the subject of tools fully here. For those who wish to learn more about tools and their use there is a book entitled *Motorcycle Workshop Practice Manual* (Book No 1454) available from the publishers of this manual.

As a general rule, it is better to buy the more expensive, good quality tools. Given reasonable use, such tools will last for a very long time, whereas the cheaper, poor quality, item will wear out faster and need to be renewed more often, thus nullifying the original saving. There is also the risk of a poor quality tool breaking while in use, causing personal injury or expensive damage to the component being worked on.

For practically all tools, a tool factor is the best source since he will have a very comprehensive range compared with the average garage or accessory shop. Having said that, accessory shops often offer excellent quality tools at discount prices, so it pays to shop around. There are plenty of tools around at reasonable prices, but always aim to purchase items which meet the relevant national safety standards. If in doubt, seek the advice of the shop proprietor or manager before making a purchase.

The basis of any toolkit is a set of spanners. While open-ended spanners with their slim jaws, are useful for working on awkwardly-positioned nuts, ring spanners have advantages in that they grip the nut far more positively. There is less risk of the spanner slipping off the nut and damaging it, for this reason alone ring spanners are to be preferred. Ideally, the home mechanic should acquire a set of each, but if expense rules this out a set of combination spanners (open-ended at one end and with a ring of the same size at the other) will provide a good compromise. Another item which is so useful it should be considered an essential requirement for any home mechanic is a set of socket spanners. These are available in a variety of drive sizes. It is recommended that the $\frac{1}{2}$-inch drive type is purchased to begin with as although bulkier and more expensive than the $\frac{3}{8}$-inch type, the larger size is far more common and will accept a greater variety of torque wrenches, extension pieces and socket sizes. The socket set should comprise sockets of sizes between 8 and 24 mm, a reversible ratchet drive, an extension bar of about 10 inches in length, a spark plug socket with a rubber insert, and a universal joint. Other attachments can be added to the set at a later date.

Maintenance and minor repair tool kit

Set of spanners 8 – 24 mm
Set of sockets and attachments
Spark plug spanner with rubber insert – 10, 12, or 14 mm as appropriate
Adjustable spanner
C-spanner/pin spanner
Torque wrench (same size drive as sockets)
Set of screwdrivers (flat blade)
Set of screwdrivers (cross-head)
Set of Allen keys 4 – 10 mm
Impact screwdriver and bits
Ball pein hammer – 2 lb
Hacksaw (junior)
Self-locking pliers – Mole grips or vice grips
Pliers – combination
Pliers – needle nose
Wire brush (small)
Soft-bristled brush
Tyre pump
Tyre pressure gauge
Tyre tread depth gauge
Oil can
Fine emery cloth
Funnel (medium size)
Drip tray
Grease gun
Set of feeler gauges
Brake bleeding kit
Strobe timing light
Continuity tester (dry battery and bulb)
Soldering iron and solder
Wire stripper or craft knife
PVC insulating tape
Assortment of split pins, nuts, bolts, and washers

Repair and overhaul toolkit

The tools in this list are virtually essential for anyone undertaking major repairs to a motorcycle and are additional to the tools listed above. Concerning Torx driver bits, Torx screws are encountered on some of the more modern machines where their use is restricted to fastening certain components inside the engine/gearbox unit. It is therefore recommended that if Torx bits cannot be borrowed from a local dealer, they are purchased individually as the need arises. They are not in regular use in the motor trade and will therefore only be available in specialist tool shops.

Plastic or rubber soft-faced mallet
Torx driver bits
Pliers – electrician's side cutters
Circlip pliers – internal (straight or right-angled tips are available)
Circlip pliers – external
Cold chisel
Centre punch
Pin punch
Scriber
Scraper (made from soft metal such as aluminium or copper)
Soft metal drift
Steel rule/straight edge
Assortment of files
Electric drill and bits
Wire brush (large)
Soft wire brush (similar to those used for cleaning suede shoes)
Sheet of plate glass
Hacksaw (large)
Valve grinding tool
Valve grinding compound (coarse and fine)
Stud extractor set (E-Z out)

Specialized tools

This is not a list of the tools made by the machine's manufacturer to carry out a specific task on a limited range of models. Occasional references are made to such tools in the text of this manual and, in general, an alternative method of carrying out the task without the manufacturer's tool is given where possible. The tools mentioned in this list are those which are not used regularly and are expensive to buy in view of their infrequent use. Where this is the case it may be possible to hire or borrow the tools against a deposit from a local dealer or tool hire shop. An alternative is for a group of friends or a motorcycle club to join in the purchase.

Valve spring compressor
Piston ring compressor
Universal bearing puller
Cylinder bore honing attachment (for electric drill)

Micrometer set
Vernier calipers
Dial gauge set
Cylinder compression gauge
Vacuum gauge set
Multimeter
Dwell meter/tachometer

Care and maintenance of tools

Whatever the quality of the tools purchased, they will last much longer if cared for. This means in practice ensuring that a tool is used for its intended purpose; for example screwdrivers should not be used as a substitute for a centre punch, or as chisels. Always remove dirt or grease and any metal particles but remember that a light film of oil will prevent rusting if the tools are infrequently used. The common tools can be kept together in a large box or tray but the more delicate, and more expensive, items should be stored separately where they cannot be damaged. When a tool is damaged or worn out, be sure to renew it immediately. It is false economy to continue to use a worn spanner or screwdriver which may slip and cause expensive damage to the component being worked on.

Fastening systems

Fasteners, basically, are nuts, bolts and screws used to hold two or more parts together. There are a few things to keep in mind when working with fasteners. Almost all of them use a locking device of some type; either a lock washer, lock nut, locking tab or thread adhesive. All threaded fasteners should be clean, straight, have undamaged threads and undamaged corners on the hexagon head where the spanner fits. Develop the habit of replacing all damaged nuts and bolts with new ones.

Rusted nuts and bolts should be treated with a rust penetrating fluid to ease removal and prevent breakage. After applying the rust penetrant, let it 'work' for a few minutes before trying to loosen the nut or bolt. Badly rusted fasteners may have to be chiselled off or removed with a special nut breaker, available at tool shops.

Flat washers and lock washers, when removed from an assembly should always be replaced exactly as removed. Replace any damaged washers with new ones. Always use a flat washer between a lock washer and any soft metal surface (such as aluminium), thin sheet metal or plastic. Special lock nuts can only be used once or twice before they lose their locking ability and must be renewed.

If a bolt or stud breaks off in an assembly, it can be drilled out and removed with a special tool called an E-Z out. Most dealer service departments and motorcycle repair shops can perform this task, as well as others (such as the repair of threaded holes that have been stripped out).

Standard torque settings

Specific torque settings will be found at the end of the specifications section of each chapter. Where no figure is given, bolts should be secured according to the table below.

Fastener type (thread diameter)	kgf m	lbf ft
5 mm bolt or nut	0.45 – 0.6	3.5 – 4.5
6 mm bolt or nut	0.8 – 1.2	6 – 9
8 mm bolt or nut	1.8 – 2.5	13 – 18
10 mm bolt or nut	3.0 – 4.0	22 – 29
12 mm bolt or nut	5.0 – 6.0	36 – 43
5 mm screw	0.35 – 0.5	2.5 – 3.6
6 mm screw	0.7 – 1.1	5 – 8
6 mm flange bolt	1.0 – 1.4	7 – 10
8 mm flange bolt	2.4 – 3.0	17 – 22
10 mm flange bolt	3.0 – 4.0	22 – 29

Chapter 1 Engine, clutch and gearbox

For modifications and information relating to later models, see Chapter 7

Contents

Specifications

Engine

	XL250 S	XR250	XL500 S	XR500
Type	Single cylinder, air-cooled, sohc four stroke			
Bore	74.0 mm	74.0 mm	89.0 mm	89.0 mm
	(2.91 in)	(2.91 in)	(3.50 in)	(3.50 in)
Stroke	57.8 mm	57.8 mm	80.0 mm	80.0 mm
	(2.73 in)	(2.73 in)	(3.15 in)	(3.15 in)
Capacity	249 cc	249 cc	498 cc	498 cc
	(15.1 cu in)	(15.1 cu in)	(30.37 cu in)	
Compression ratio	9.1:1	9.6:1	8.6:1	8.6:1
Power output	20 bhp @	24 bhp @	32 bhp @	36 bhp @
	7500 rpm	9000 rpm	6250 rpm	6500 rpm
Maximum torque	2.12 kg m	2.1 kg m	4.14 kg m	4.2 kg m
	(15.3 ft lb)	(15.2 ft lb)	(29.9 ft lb)	(30.6 ft lb)
	@ 6000 rpm	@ 7000 rpm	@ 5000 rpm	@ 5500 rpm

	XL models	XR models
Compression pressure		
At cranking speed	12.5 kg cm² (175 psi)	13.5 ± 1.5 kg cm² (192 ± 21 psi)

Camshaft and rockers

	XL models	XR models
Cam lift:		
Inlet	36.362 mm (1.4316 in)	36.789 mm (1.4452 in)
Wear limit	36.30 mm (1.429 in)	36.65 mm (1.443 in)
Exhaust	36.256 mm (1.4274 in)	36.605 mm (1.4411 in)
Wear limit	36.20 mm (1.425 in)	36.55 mm (1.443 in)

	All models
Camshaft journal outside diameter:	
Right	23.954 – 23.975 mm (0.9431 – 0.9439 in)
Wear limit	23.9 mm (0.94 in)
Left	19.954 – 19.975 mm (0.9856 – 0.9864 in)
Wear limit	19.9 mm (0.78 in)
Rocker arm bore	12.00 – 12.018 mm (0.4724 – 0.4731 in)
Wear limit	12.05 mm (0.474 in)
Rocker spindle diameter	11.966 – 11.984 mm (0.4711 – 0.4718 in)
Wear limit	11.91 mm (0.469 in)

Valves and valve guides

	250 models	500 models
Valve spring free length:		
Inner	43.6 mm (1.72 in)	38.1 mm (1.50 in)
Wear limit	42.5 mm (1.67 in)	37.0 mm (1.46 in)
Outer	35.58 mm (1.40 in)	36.24 mm (1.43 in)
Wear limit	34.5 mm (1.36 in)	35.3 mm (1.39 in)
Valve stem diameter:		
Inlet	5.475 – 5.490 mm (0.2037 – 0.2161 in)	6.575 – 6.590 mm (0.2589 – 0.2594 in)
Wear limit	5.465 mm (0.2152 in)	6.565 mm (0.2585 in)
Exhaust	5.455 – 5.470 mm (0.2148 – 0.2154 in)	6.560 – 6.570 mm (0.2583 – 0.2587 in)
Wear limit	5.445 mm (0.2144 in)	6.550 mm (0.2579 in)
Valve guide bore:		
Inlet	5.500 – 5.512 mm (0.2165 – 0.2170 in)	6.600 – 6.615 mm (0.2598 – 0.2421 in)
Wear limit	5.53 mm (0.218 in)	6.63 mm (0.261 in)
Exhaust	5.500 – 5.512 mm (0.2165 – 0.2170 in)	6.600 – 6.615 mm (0.2598 – 0.2421 in)
Wear limit	5.53 mm (0.218 in)	6.63 mm (0.261 in)
Stem/guide clearance:		
Inlet	0.010 – 0.047 mm (0.0004 – 0.0019 in)	0.010 – 0.040 mm (0.0004 – 0.0016 in)
Wear limit	0.06 mm (0.0024 in)	0.65 mm (0.0026 in)
Exhaust	0.030 – 0.057 mm (0.0012 – 0.0022 in)	0.030 – 0.055 mm (0.0012 – 0.0022 in)
Wear limit	0.07 mm (0.0028 in)	0.080 mm (0.0031 in)
Valve face width	1.2 – 1.4 mm (0.048 – 0.055 in)	1.2 – 1.4 mm (0.048 – 0.055 in)
Wear limit	2.0 mm (0.08 in)	2.0 mm (0.08 in)
Valve seat width	1.2 – 1.4 mm (0.048 – 0.055 in)	1.2 – 1.4 mm (0.048 – 0.055 in)
Wear limit	2.0 mm (0.08 in)	2.0 mm (0.08 in)
Camshaft bearing bore:		
Left	20.000 – 20.021 mm (0.7874 – 0.7882 in)	20.000 – 20.021 mm (0.7874 – 0.7882 in)
Wear limit	20.05 mm (0.789 in)	20.07 mm (0.790 in)
Right	24.000 – 24.021 mm (0.9449 – 0.9457 in)	24.000 – 24.021 mm (0.9449 – 0.9457 in)
Wear limit	24.05 (0.947 in)	24.07 mm (0.948 in)

Valve timing

	XL250 At 1mm/0 lift	XR250 At 0 lift	500 models At 1mm/0 lift
Inlet opens	5°/58° BTDC	10° BTDC	5°/58° BTDC
Inlet closes	30°/96° ABDC	40° ABDC	40°/106° ABDC
Exhaust opens	35°/83° BBDC	10° BBDC	45°/95° BBDC
Exhaust closes	5°/65° ATDC	40° ATDC	5°/65° ATDC

Valve clearances (cold engine)

Inlet .... 0.05 mm (0.002 in) 0.08 mm (0.003 in) 0.05 mm (0.002 in)

Exhaust .... 0.10 mm (0.004 in) 0.10 mm (0.004 in) 0.10 mm (0.004 in)

Cylinder barrel and piston

	250 models	500 models
Standard bore size	74.00 – 74.01 mm (2.913 – 2.914 in)	88.00 – 89.01 mm (3.5039 – 3.5041 in)
Wear limit	74.11 mm (2.918 in)	89.11 mm (3.508 in)
Taper limit	0.05 mm (0.002 in)	0.05 mm (0.002 in)
Ovality limit	0.05 mm (0.002 in)	0.05 mm (0.002 in)
Piston diameter (at skirt)	73.97 – 73.99 mm (2.912 – 2.913 in)	88.97 – 88.99 mm (3.5027 – 3.5034)
Wear limit	73.88 mm (2.909 in)	88.88 mm (3.499 in)
Gudgeon pin bore	19.002 – 19.008 mm (07481 – 0.7483 in)	21.002 to 21.008 mm (0.8268 – 0.8271 in)
Wear limit	19.08 mm (0.751 in)	21.08 mm (0.830 in)
Gudgeon pin diameter	18.994 – 19.000 mm (0.7478 – 0.7480 in)	20.994 – 21.000 mm (0.8265 – 0.8268 in)
Wear limit	18.96 mm (0.747 in)	20.96 mm (0.825 in)

Piston ring end gap:

	250 models	500 models
Top/2nd	0.15 – 0.35 mm (0.006 – 0.010 in)	0.30 – 0.50 mm (0.0118 – 0.0197 in)
Wear limit	0.5 mm (0.020 in)	0.65 mm (0.026 in)
Oil (side rails, max)	0.2 – 0.9 mm (0.007 – 0.035 in)	0.2 – 0.9 mm (0.007 – 0.035 in)

Piston ring to groove clearance:

	250 models	500 models
Top	0.015 – 0.045 mm (0.0006 – 0.0018 in)	0.030 – 0.65 mm (0.0012 – 0.0026 in)
2nd	0.015 – 0.045 mm (0.0006 – 0.0018 in)	0.015 – 0.045 mm (0.0006 – 0.0018 in)
Piston to cylinder clearance	0.01 – 0.04 mm (0.0004 – 0.0006 in)	0.01 – 0.04 mm (0.0004 – 0.0006 in)
Wear limit	0.1 mm (0.004 in)	0.1 mm (0.004 in)

Crankshaft

All models

Big-end axial clearance:

250 models .... 0.05 – 0.45 mm (0.002 – 0.017 in)

Wear limit .... 0.6 mm (0.024 in)

500 models .... 0.05 – 0.65 mm (0.0020 – 0.0256 in)

Wear limit .... 0.8 mm (0.031 in)

Big-end radial clearance .... 0.006 – 0.018 mm (0.0002 – 0.0007 in)

Wear limit .... 0.05 mm (0.002 in)

Small-end bearing ID:

250 models .... 19.020 – 19.041 mm (0.7488 – 0.7496 in)

Wear limit .... 19.07 mm (0.751 in)

500 models .... 21.020 – 21.041 mm (0.8276 – 0.8284 in)

Wear limit .... 21.07 mm (0.830 in)

Crankshaft runout (max) .... 0.1 mm (0.004 in)

Balancer assembly

Front balancer holder O.D. .... 39.964 – 39.980 mm (1.5734 – 1.5740 in)

Wear limit .... 39.91 mm (1.571 in)

Front balancer holder I.D. .... 26.007 – 26.020 mm (1.0239 – 1.0244 in)

Wear limit .... 26.05 mm (1.026 in)

Rear balancer I.D. .... 26.007 – 26.020 mm (1.0239 – 1.0244 in)

Wear limit .... 26.05 mm (1.026 in)

Clutch

Free play (measured at handlebar lever end) .... 15 – 25 mm ($\frac{5}{8}$ – 1 in)

Clutch spring free length:

250 models .... 37.3 mm (1.46 in)

Wear limit .... 35.8 mm (1.41 in)

500 models .... 41.0 mm (1.61 in)

Wear limit .... 39.5 mm (1.56 in)

Friction plate thickness .... 2.62 – 2.78 mm (0.102 – 0.109 in)

Wear limit .... 2.3 mm (0.091 in)

Plain plate warpage (max) .... 0.3 mm (0.012 in)

Gearbox

	250 models	500 models
Type		Five speed, constant mesh
Ratios:		
1st	2.800:1 (42/15T)	2.462:1 (32/13T)

2nd	1.850:1 (37/20T)	1.647:1 (28/17T)
3rd	1.375:1 (33/24T)	1.250:1 (25/20T)
4th	1.111:1 (37/27T)	1.000:1 (23/23T)
5th	0.900:1 (27/30T)	0.840:1 (21/25T)
Gear backlash:		
1st and 2nd	N/A	0.044 – 0.133 mm (0.0017 – 0.0052 in)
Wear limit	0.20 mm (0.008 in)	0.20 mm (0.008 in)
3rd, 4th and 5th	N/A	0.046 – 0.140 mm (0.0018 – 0.0055 in)
Wear limit	0.20 mm (0.008 in)	0.20 mm (0.008 in)
Gear side clearance:		
Mainshaft 4th	N/A	0.06 – 0.41 mm (0.002 – 0.016 in)
Wear limit	N/A	0.50 mm (0.020 in)
Mainshaft 5th	N/A	0.05 – 0.58 mm (0.002 – 0.023 in)
Wear limit	N/A	0.65 mm (0.026 in)
Layshaft 1st	N/A	0.05 – 0.22 mm (0.002 – 0.009 in)
Wear limit	N/A	0.30 mm (0.012 in)
Layshaft 3rd	N/A	0.06 – 0.41 mm (0.002 – 0.016 in)
Wear limit	N/A	0.50 mm (0.020 in)
Gear dog clearance:	**All models**	
In neutral, max	0.3 mm (0.01 in)	
Selector fork bore diameter:		
Centre fork	12.000 – 12.021 mm (0.4724 – 0.4733 in)	
Wear limit	12.05 mm (0.474 in)	
Right and left forks	15.000 – 15.021 mm (0.5906 – 0.5914 in)	
Wear limit	15.05 mm (0.593 in)	
Selector fork shaft diameter:		
Centre fork	11.966 – 11.984 mm (0.4711 – 0.4718 in)	
Wear limit	11.91 mm (0.496 in)	
Right and left forks	14.966 – 14.984 mm (0.5892 – 0.5899 in)	
Wear limit	14.91 mm (0.587 in)	
Selector fork width	4.93 – 5.00 mm (0.194 – 0.197 in)	
Wear limit	4.50 mm (0.18 in)	

Primary transmission

Type	Gear
Reduction ratio	2.379:1 (69/29T)

Secondary transmission

Type	Chain
Reduction ratio:	3.785:1 (53/14T)
XL250 S	3.785:1 (53/14T)
XR250	4.076:1 (53/13T)
XL500 S	2.786:1 (39/14T)
XR500	3.429:1 (48/14T)

Torque settings

	lbf ft	kgf m
Cylinder head cover	7 – 10	1.0 – 1.4
Cylinder head:		
250 models	25 – 29	3.5 – 4.0
500 models	16 – 20	2.2 – 2.8
Clutch centre nut	33 – 43	4.5 – 6.0
Balancer holder bracket	16 – 20	2.2 – 2.8
Ignition reluctor	33 – 43	4.5 – 6.0
Alternator rotor:		
250 models	69 – 76	9.5 – 10.5
500 models	72 – 87	10.0 – 12.0
Camshaft sprocket bolts	12 – 16	1.7 – 2.3
Crankcase (8 x 1.25 mm)	16 – 20	2.2 – 2.8
Crankcase (6 x 1.00 mm)	7 – 10	1.0 – 1.4
Crankcase (9 x 1.25 mm)		
250 models	N/A	N/A
500 models	20 – 24	2.7 – 3.3
Crankcase (10 x 1.25 mm):		
250 models	N/A	N/A
500 models	23 – 27	3.2 – 3.8

1   General description

The engine unit used on the Honda XL and XR models covered by this manual is essentially similar in construction with minor variations relating to the machine's intended purpose. In each case the engine, primary drive, clutch and gearbox are built as a single unit and share common casings. The engine is an air cooled, single cylinder four-stroke, employing a single overhead camshaft which operates the paired inlet and exhaust valves. The four-valve design has been chosen in the interests of lower reciprocating weight and to permit a greater valve area for a given bore size.

A conventional crankshaft assembly is used, featuring full-circle flywheels, a caged roller big-end bearing and journal ball main bearings. To reduce the effects of vibration, a pair of balancer weights are fitted, these being driven by a single-row chain from the crankshaft. The front balancer weight runs on a shaft supported in an eccentric holder forward of the crankshaft assmbly. The rear balancer weight is mounted on the left-hand end of the gearbox mainshaft, thus minimising the extra weight and bulk incurred by the adoption of a balancer system.

Power from the crankshaft is fed by gear to the wet multiplate clutch and thence to the five-speed constant-mesh gearbox. The crankcase casings are arranged to split horizontally, the crankcase proper being formed as a separate chamber with the primary drive and gearbox assembly sharing the remaining area.

Lubrication is of the wet sump type, the oil reservoir being formed in the bottom of the crankcase. Oil is picked up by the trochoid oil pump and fed directly to the main engine and transmission components.

2   Operations with the engine/gearbox unit in the frame

The following items can be overhauled with the engine/gearbox unit installed in the frame:

XL250 S and XR250 only
1   Cylinder head and valves
2   Camshaft
3   Cylinder barrel and piston

All models
4   Clutch and primary drive
5   Kickstart mechanism
6   Ignition pickup
7   Gear selector mechanism
8   Alternator
9   Final drive sprocket

When several operations need to be undertaken simultaneously, it would probably be an advantage to remove the complete unit from the frame, a comparatively simple operation that should take approximately one hour. This will give the advantage of battery access and more working space.

3   Operations with the engine/gearbox unit removed from the frame

It will be necessary to remove the engine/gearbox unit from the frame to gain access to the following:

XL500S and XR500 only
1   Cylinder head and valves
2   Camshaft and cam chain
3   Cylinder barrel and piston

All models
4   Crankshaft
5   Balancer and drive assembly
6   Gearbox components
7   Oil pump

4   Removing the engine/gearbox unit from the frame

1   Before commencing any dismantling work it will be necessary to drain the engine oil, preferably whilst the engine is warm. If possible, leave the oil to drain overnight before commencing work in the morning. It is helpful to remove the skid plate at this stage. The plate is retained by two bolts passing through lugs at the rear of the plate and by a single bolt at the front (single domed nut, XL250S model). With the machine on its proper stand, place a drain tray of about 2.0 litres/4.0 pints below the left-hand side of the crankcase, remove the drain plug and leave the oil to drain completely. Once draining is complete, refit the plug finger tight to preclude its loss.

2   It will be necessary to arrange the machine so that it is supported securely. If possible a stand or wooden crate should be positioned beneath the frame, to the rear of the engine unit and webbing tie-down straps used to secure the machine in a vertical position. Failing this, the machine can be supported by the side stand, given that reasonable care is exercised to avoid its rolling forward and off the stand. To this end, turn the steering fully to the left and place chocks behind the front and rear wheels. It is helpful, though not essential, to place the machine on a raised working platform or bench.

3   The seat is retained by a bolt on either side, immediately below the lower edge of the seat. Release the two bolts, and where fitted, the seat strap. This is secured by a small bolt at either end. The bolts also retain the rear suspension units and should be refitted once the seat strap has been removed. The seat can now be lifted clear to reveal the single fuel tank fixing bolt. Check that the fuel tap is turned to the 'Off' position, and disconnect the fuel pipe. Slacken and remove the tank fixing bolt, then pull the rear of the tank up and back to disengage the mounting rubbers at the front. Place the tank and seat to one side to await reassembly, bearing in mind the fire risk posed by the fuel tank.

4   On XL models, disconnect and remove the battery to prevent any risk of short circuits which may occur during subsequent dismantling. Trace the ignition pulser and alternator leads up from the engine, releasing the relevant wiring ties, and disconnect them at the connector blocks beneath the top frame tubes. Coil the leads carefully and place them on the upper crankcase to avoid snagging during removal. Pull off the sparking plug cap and lodge it and the HT cable clear of the cylinder head.

5   Slacken and remove the nut which retains the clutch cable anchor bracket at the lower end of the cable. Free the bracket and temporarily replace the nut, then disengage the cable from the clutch arm. Lodge the cable and bracket around the top frame tube, displacing it from the guide clip at the cylinder head.

6   Pull off the right-hand side panel, if this is still in position. Remove the two screws which retain the exhaust heat shield and lift it away. Slacken the clamp at the junction of the exhaust pipes and silencer. Remove the two nuts which retain each of the exhaust pipe flanges. The exhaust pipes can now be pulled forward to clear the exhaust ports and silencer and lifted away from the frame.

7   Slacken the hose clips which retain the carburetor to the induction flange and the air filler adaptor. The carburettor can now be pulled clear of the rubber adaptors. Free the drain hose from the guide clip next to the swinging arm. It is not essential to release the throttle cables from the carburettor unless attention to the latter is necessary. If left connected, the cables and the carburettor can now be lodged on the top frame tubes, clear of the engine unit.

8   Remove the two sprocket guard retaining screws and lift the guard away. Release the two sprocket mounting bolts having first bent back the locking tubes. There will normally be sufficient slack in the drive chain to allow it to be pulled to one side as the sprocket is slid off its splines. If this is not the case, slacken the rear wheel spindle and adjusters to allow the wheel to be pushed forward. The chain can be left around the swinging

arm pivot tube with no adverse effects.

9   Slacken and remove the rear brake cable adjuster nut, then remove the right-hand footrest/brake pedal assembly together with the cable. Note that the rear brake switch spring should be disconnected as the assembly is removed. The left-hand footrest assembly should now be released in a similar fashion to permit removal of the engine lower mounting bolt. Pull off the crankcase breather hose from its stub on the upper crankcase half.

10   Dismantle the engine front mounting plate by releasing the four securing bolts, then dismantle the cylinder head mountings. The engine unit is now supported by the upper and lower mounting bolts alone, and will require support as these are removed. It was found expedient to wedge a short length of wood betwen the frame and cylinder barrel to support the engine unit during the removal of the latter. Once all of the mounting bolts are removed, grasp the engine and allow the lower half to move forward to clear the frame at the rear. The unit can now be lowered away from the frame. It is helpful to have available an assistant at this latter stage. This however, is by no means essential in view of the unit's relatively light weight.

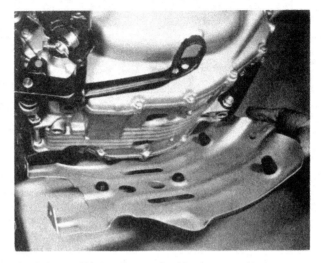

4.1 Release skid plate from underside of crankcase

4.3a Remove seat mounting bolts...

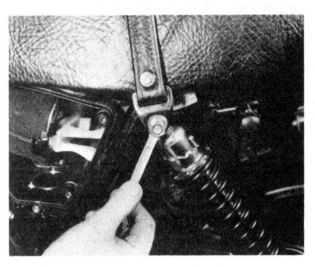

4.3b ...and release seat strap (where fitted)

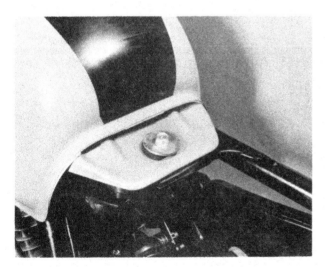

4.3c Tank is retained by single mounting bolt

4.3d Lift tank and pull rearwards to remove

4.4 Trace and disconnect electrical connections

4.5 Free clutch cable and bracket (arrowed)

4.6 Slacken silencer clamp bolt

4.9a Footrest/brake pedal assembly is secured by bolt

4.9b LH footrest is secured in similar manner

4.10a Dismantle engine mounting plate

4.10b Engine unit can now be lifted clear of frame

5 Dismantling the engine and gearbox: general

1 Before commencing work on the engine unit, the external surfaces must be cleaned thoroughly. A motor cycle engine has very little protection from road grit and othe foreign matter, which will sooner or later find its way into the dismantled engine if this simple precaution is not observed.

2 One of the proprietary engine cleaning compounds such as 'Gunk' or 'Jizer' can be used to good effect, especially if the compound is allowed to penetrate the film of oil and grease before it is washed away. When washing down, make sure that water cannot enter the inlet or exhaust ports or the electrical system, particularly if these parts are now exposed.

3 Never use force to remove any stubborn part, unless mention is made of this requirement in the text. There is invariably good reason why a part is difficult to remove, often because the dismantling operation has been tackled in the wrong sequence.

4 The engine units employed in the Honda XL and XR models are fairly straightforward in construction and pose few specific dismantling problems. The only special tool that will prove essential is an extractor for the alternator rotor. On many machines, another component from the chassis can be used as an extractor bolt, the rear wheel spindle usually proving to be ideal for this purpose. This is not true of the above models where an unusually large extractor thread is provided. If a 22 mm thread bolt is available, it can be used to good effect, but failing this the correct Honda tool, a multi-purpose extractor, will be required. Its part number is 07733-0020001.

6 Dismantling the engine/gearbox unit: removing the cylinder head cover, camshaft and cylinder head

1 As mentioned earlier in this Chapter, on the 250 models cylinder head removal can be carried out with the engine unit in or out of the frame. In the case of the 500 models, the height of the engine and the rather confined frame loop require engine removal as a precursor to this operation. If the work is to be undertaken with the engine unit installed (250 models), it will first be necessary to remove the following items.
1 Seat and fuel tank
2 Cylinder head mounting plates
3 Carburettor
4 Exhaust pipes
Full details describing the removal of the above will be found in Section 4 of this Chapter.

2 Slacken the locknuts at the upper and lower ends of the decompressor cable. Displace the lower adjuster from its locating bracket and free the lower end of the cable from the operating lever. Repeat the procedure to release the upper end of the cable. Remove the two bolts which secure each of the value adjuster inspection covers and lift them away.

3 The cylinder head cover is retained by a total of 12 bolts, two of which are located inside the valve adjuster inspection holes. Note that two of the cylinder head holding bolts (domed nuts, in the case of the 500 models) are located on the left-hand side of the cylinder head, recessed below the flat area surrounding the sparking plug. These should not be disturbed at this stage. Slacken the cylinder head bolts evenly, in a diagonal sequence to prevent warping. The cylinder head cover can now be lifted away, together with the valve rockers which are supported inside the cover. Remove the rubber end plug if this remains in the recess in the cylinder head.

4 Remove the larger of the two inspection plugs from the engine left-hand outer casing so that the crankshaft can be rotated by passing a socket through the inspection hole and onto the alternator rotor securing bolt. The camshaft sprocket bolts should now be removed, turning the crankshaft to gain access to each bolt as required. Arrange the camshaft sprocket so that one of the bolt holes is uppermost with the two cutouts symmetrically disposed in relation to the gasket face. Displace the sprocket to the right until it drops clear of its locating shoulder. The cam chain can now be slid off to the left of the sprocket. Wrap some wire around the chain to prevent it from falling into the cam chain tunnel, then manoeuvre the camshaft and sprocket clear of the cylinder head.

5 Release the two cylinder head nuts. These are located on the front and rear faces of the cylinder barrel, and retain studs which project down from the cylinder head. The cylinder head is now retained by a total of four bolts in the case of the 250 models, or by four domed nuts and washers on the 500 versions. The nuts or bolts should be slackened progressively, by about one flat at a time, in a diagonal sequence. This will ensure that the loading on the head casting remains even, avoiding any tendency towards warpage.

6 Slacken and remove the upper tensioner bolt and O-ring. The cylinder head can now be removed. It is quite likely that the head will be stuck to the cylinder barrel by the gasket, and if this proves to be the case it may be necessary to break the joint by tapping around the head with a soft-faced mallet. Care must be taken to avoid damage to the cooling fins. These are rather brittle and are easily chipped. For obvious reasons, no attempt should be made to prise apart the joint by levering between the fins.

6.3a Note that two cylinder cover bolts are inside inspection covers

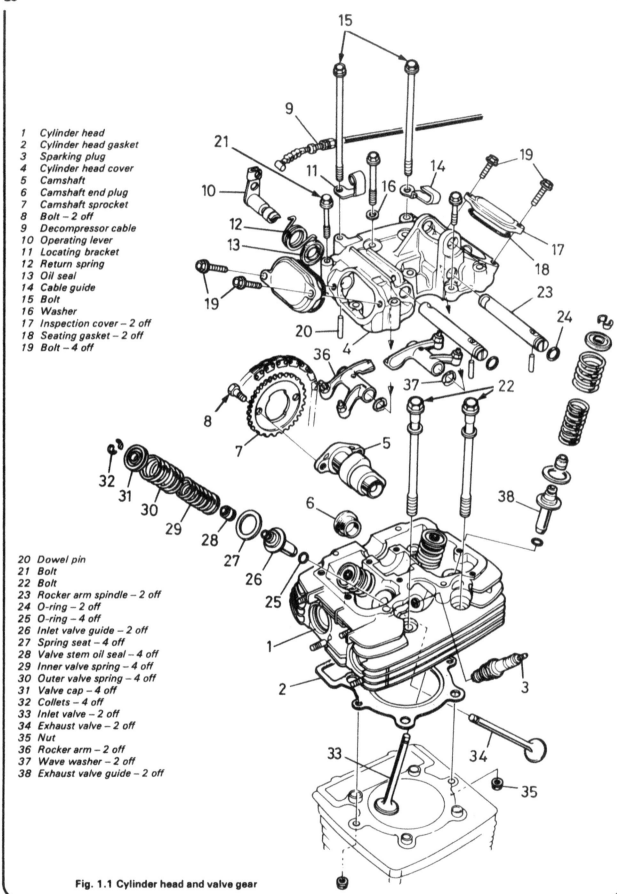

1 Cylinder head
2 Cylinder head gasket
3 Sparking plug
4 Cylinder head cover
5 Camshaft
6 Camshaft end plug
7 Camshaft sprocket
8 Bolt – 2 off
9 Decompressor cable
10 Operating lever
11 Locating bracket
12 Return spring
13 Oil seal
14 Cable guide
15 Bolt
16 Washer
17 Inspection cover – 2 off
18 Seating gasket – 2 off
19 Bolt – 4 off

20 Dowel pin
21 Bolt
22 Bolt
23 Rocker arm spindle – 2 off
24 O-ring – 2 off
25 O-ring – 4 off
26 Inlet valve guide – 2 off
27 Spring seat – 4 off
28 Valve stem oil seal – 4 off
29 Inner valve spring – 4 off
30 Outer valve spring – 4 off
31 Valve cap – 4 off
32 Collets – 4 off
33 Inlet valve – 2 off
34 Exhaust valve – 2 off
35 Nut
36 Rocker arm – 2 off
37 Wave washer – 2 off
38 Exhaust valve guide – 2 off

Fig. 1.1 Cylinder head and valve gear

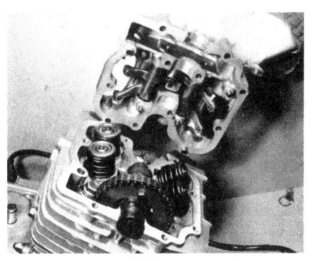

6.3b Remove securing bolts and lift cover away

6.4a Slacken and remove camshaft sprocket bolts

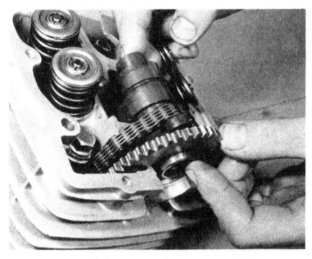

6.4b Manoeuvre camshaft clear of sprocket and chain

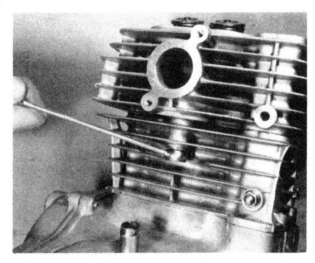

6.5 Do not overlook bolts on underside of cylinder head

6.6 Remove upper cam chain tensioner bolt (or nut)

7 Dismantling the engine/gearbox unit: removing the cylinder barrel and piston

1   Removal of the cylinder barrel and piston can be accomplished after the cylinder head and camshaft have been removed, as described in the preceding section. Note that in the case of the 500 models it will be necessary to remove the engine unit from the frame.

2   Remove the remaining cam chain tensioner nut, washer and O-ring (250 models) or nut and sealing washer (500 models) and move the tensioner assembly clear of the cam chain tunnel. The cam chain guide on the exhaust side of the barrel can be lifted away. Remove the cylinder barrel retaining nuts and bolts. The cylinder barrel can now be removed by pulling it upwards. If necessary, the barrel to crankcase joint can be freed by tapping around it with a soft-faced mallet or by rocking the cylinder barrel to and fro.

3   Lift the barrel upwards by about two inches, then pack some clean rag around the crankcase mouth to catch any debris or portions of broken piston ring which might otherwise drop into the crankcase as the barrel is pulled clear of the piston. The barrel can now be removed completely, taking care to support

the connecting rod and piston as the latter emerges from the bore.

4   Prise out one of the gudgeon pin circlips using a small electrical screwdriver or a similar tool in the slot provided. Support the piston and push the gudgeon pin out until the connecting rod is freed. The piston can now be lifted away. If the pin is a tight fit, it may be necessary to warm the piston so that the grip on the gudgeon pin is released. A rag soaked in warm water will suffice, if it is placed on the piston crown. The piston may be lifted from the connecting rod once the gudgeon pin is clear of the small-end eye.

5   If the gudgeon pin is still a tight fit after warming the piston it can be lightly tapped out of position with a hammer and soft metal drift. **Do not** use excess force and make sure the connecting rod is supported during this operation, or there is a risk of its bending.

6   When the piston is free of the connecting rod, remove the gudgeon pin completely, by removing the scond clip. Place the piston, rings and gudgeon pin aside for further attention, but discard the circlips. They should never be re-used; new circlips must be obtained and fitted during rebuilding.

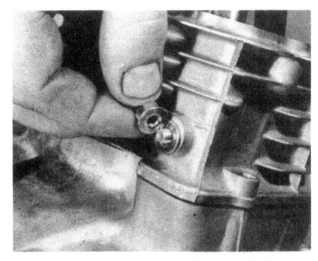

7.2a Release lower tensioner bolt washer and O-ring...

7.2b ...to allow tensioner to be displaced

7.2c Slacken cylinder barrel base bolts

7.2d Cylinder barrel can now be lifted away

7.4 Pack crankcase mouth with rag, then displace circlips

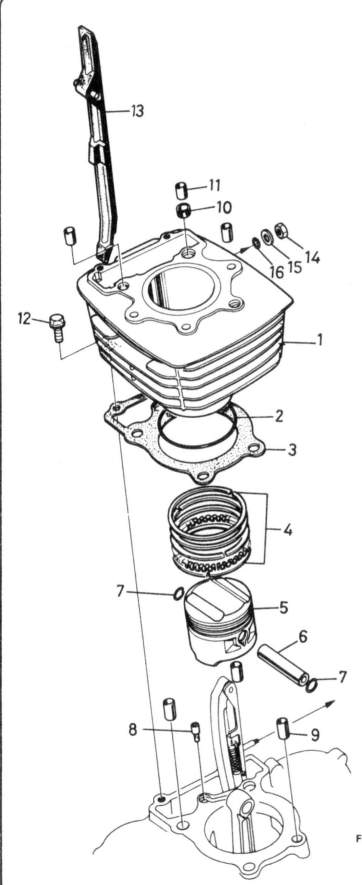

1 Cylinder barrel
2 O-ring
3 Cylinder base gasket
4 Piston ring set
5 Piston
6 Gudgeon pin
7 Circlip – 2 off
8 Oil jet (restrictor)
9 Hollow dowel – 2 off
10 Seal
11 Hollow dowel – 3 off
12 Bolt
13 Static chain tensioner blade
14 Nut
15 Washer
16 O-ring

Fig. 1.2 Cylinder barrel and piston – 250 models

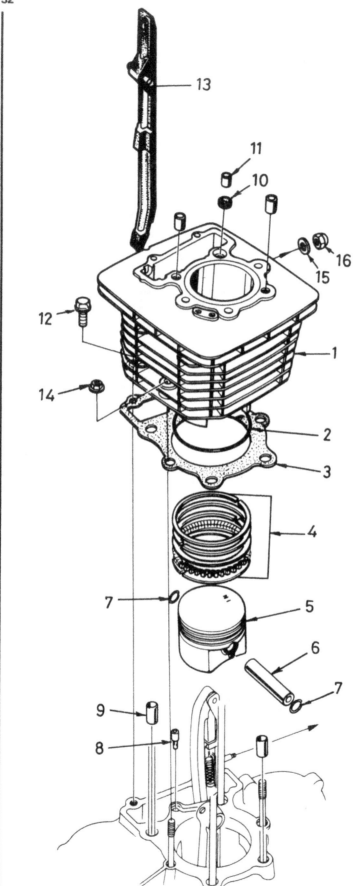

1 Cylinder barrel
2 O-ring
3 Cylinder base gasket
4 Piston ring set
5 Piston
6 Gudgeon pin
7 Circlip – 2 off
8 Oil jet (restrictor)
9 Hollow dowel – 2 off
10 Seal
11 Hollow dowel – 3 off
12 Bolt
13 Static chain tensioner blade
14 Nut
15 Washer
16 Nut

Fig. 1.3 Cylinder barrel and piston – 500 models

8 Dismantling the engine/gearbox unit: removing the right-hand outer casing and ignition pickup

1 The right-hand outer casing can be removed with the engine unit in or out of the frame. In the former instance it will be necessary to carry out the following operations prior to removal:

1 Remove the skid plate
2 Drain the engine/transmission oil
3 Remove the kickstart lever
4 Remove the right-hand footrest
5 Disconnect the rear brake lamp switch spring
6 Remove the rear brake pedal
7 Disconnect the clutch and decompressor cables

The various tasks mentioned above are described in Sections 4 and 6 of this Chapter.

2 Slacken and remove the hexagon-headed screws from the periphery of the casing, then lift it away, taking care not to damage the kickstart shaft oil seal as it passes over the shaft splines. It is possible that a small amount of residual oil will be released as the casing is lifted clear, and some provision should be made to catch this.

3 The ignition pickup coil, or pulser coil, will remain with the casing, and should be left undisturbed unless renewal is required. The ignition rotor assembly incorporates a centrifugal advance mechanism – automatic timing unit (ATU) – and is located on the crankshaft end. It can be removed after releasing the retaining nut and washer. Note that the ATU is located by a small pin in the crankshaft end. This should be removed if it is loose to preclude its loss, as should the oil feed quill and spring in the crankshaft end.

9 Dismantling the engine/gearbox unit: removing the clutch and primary drive pinion

1 The clutch and primary drive pinion can be removed with the engine in or out of the frame. Where the engine is installed a certain amount of preliminary dismantling will be required and in either case it will be necessary to remove the right-hand outer casing. Details of both will be found in Section 8 of this Chapter.

2 Remove the pushrod and thrust bearing from the centre of the clutch release plate. Slacken the four bolts which retain the release plate and lift it away. If available, the Honda clutch holding tool, part number 07923-4280000 should be fitted to the clutch centre whilst the central retaining nut is removed. Failing this, the clutch can be locked as described below. Obtain two flat washers and temporarily fit them and two clutch bolts to compress two opposing clutch springs. The clutch will now be under tension, but the clutch centre nut should still be accessible. It will now be necessary to prevent the clutch from turning as the nut is slackened. This can be done by locking the crankshaft. If the cylinder head, barrel and piston have been removed, place a smooth round metal bar through the connecting rod small-end eye, resting the ends of the bar on suitably positioned wooden strips. The wooden strips or blocks are essential to prevent the crankcase gasket from becoming marked.

4 An alternative to the above is to use a strap wrench fitted around the alternator rotor, having first removed the left-hand casing. The clutch can also be locked via the gearbox, by selecting top gear and restraining the gearbox sprocket. If the unit is in the frame this can be done by applying the rear brake, thus locking the entire drive train. If, however, the unit is on the workbench it will be necessary to contrive some means of holding the gearbox sprocket. The simplest method is to wrap a length of final drive chain around the sprocket and secure the ends with a self locking wrench.

5 Once the clutch assembly has been locked in position, the central retaining nut can be removed together with the Belville washer which locks it. The clutch assembly can now be pulled off its splines and placed to one side to await further examination. The primary drive pinion is located by splines on the crankshaft end and may be slid off once the automatic timing unit has been removed. The latter is retained by a nut and washer and should have been removed as described in the previous Section.

10 Dismantling the engine/gearbox unit: removing the oil pump

1 The oil pump can be removed after the clutch has been dismantled as described in the preceding Section. Start by removing the retainer plate which retains the kickstart idler gear. The plate is secured by two bolts which also pass through and retain the pump body. Lift the plate away, followed by the idler gear. The pump assembly is now clear and can be removed from the crankcase.

11 Dismantling the engine/gearbox unit: removing the left-hand outer casing and alternator assembly

1 The alternator assembly can be removed with the engine on the workbench, or installed in the frame. In the latter instance the following operations must be carried out, using the preceding sections as references:

1 Remove fuel tank and seat
2 Disconnect alternator output leads
3 Remove skid plate and gearchange pedal
4 Remove sprocket cover
5 Drain engine oil

2 Slacken and remove the hexagon-headed screws which retain the outer casing to the crankcase. The cover can now be lifted away complete with the alternator stator assembly which is mounted on its inside face. Note that unless the stator components require renewal the assembly should be left undisturbed.

3 The rotor is mounted on the tapered end of the crankshaft. It is located by a Woodruff key and is secured by a central flanged bolt. On the machine featured in this manual the bolt was found to be extremely tight and this is likely to prove a common occurence, particularly where the rotor is being removed for the first time since initial assembly. A secure method of preventing the rotation of the crankshaft will therefore be necessary.

4 If the engine is in the frame rotation can best be prevented by selecting top gear and locking the crankshaft through the drive chain by applying the rear brake. An alternative would be to use a stout strap wrench around the inner edge of the rotor. Where the unit is being stripped prior to crankcase separation, the crankshaft can be immobilised by passing a smooth round bar through the connecting rod small-end eye and supporting the projecting ends on wooden blocks placed against the crankcase mouth. With the crankshaft restrained by one of the above methods, the securing bolt can be removed.

5 As mentioned earlier in this Chapter, it will be necessary to obtain a Honda service tool, part number 07733-0020001, to remove the flywheel rotor which will prove to be a tight fit on its taper. The only realistic alternative to the correct service tool would be a 22 mm fine pitch bolt, but this is unlikely to be found in most workshops. It should be noted that a legged puller should **not** be used because the rotor will almost certainly be damaged. Fit the extractor or bolt, screwing it inwards to jack the rotor clear of the crankshaft. If the rotor resists removal do not continue tightening the extractor bolt down, because damage may result. Tighten the bolt fully and then strike the bolt head sharply with a hammer. This action will usually have the desired effect in breaking the fit between the tapers.

8.2 Remove screws and lift right-hand casing away

8.3a Unscrew retaining nut and displace pin...

8.3b ...to free oil feed quill and spring

9.3 Use washers and springs to lock clutch unit

10.1 Remove retainer, idler pinion and oil pump

11.2 Release LH cover complete with alternator stator

11.4 Lock crankshaft as shown to release rotor bolt

11.5 Honda extractor is used to draw rotor off

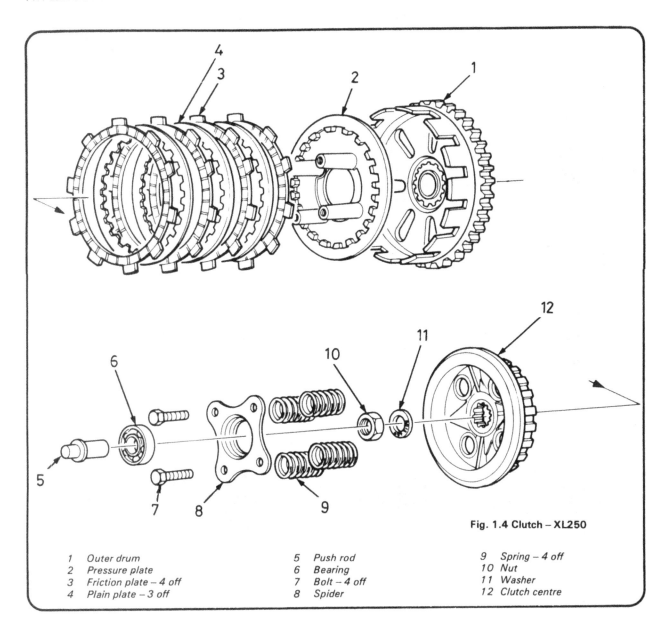

Fig. 1.4 Clutch – XL250

1	Outer drum	5	Push rod	9	Spring – 4 off
2	Pressure plate	6	Bearing	10	Nut
3	Friction plate – 4 off	7	Bolt – 4 off	11	Washer
4	Plain plate – 3 off	8	Spider	12	Clutch centre

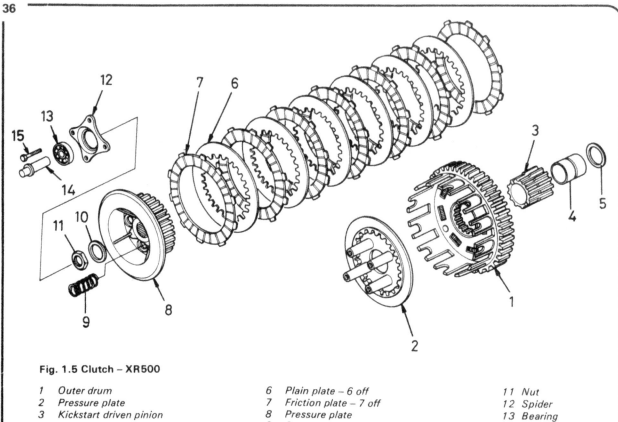

Fig. 1.5 Clutch – XR500

1	Outer drum	6 Plain plate – 6 off	11 Nut
2	Pressure plate	7 Friction plate – 7 off	12 Spider
3	Kickstart driven pinion	8 Pressure plate	13 Bearing
4	Bush (sleeve)	9 Spring – 4 off	14 Push rod
5	Thrust washer	10 Washer	15 Bolt – 4 off

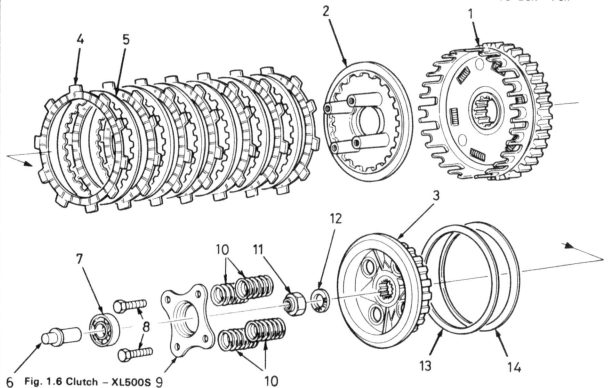

Fig. 1.6 Clutch – XL500S

1	Outer drum	6 Push rod	11 Nut
2	Pressure plate	7 Bearing	12 Washer
3	Clutch centre	8 Bolt – 4 off	13 Anti judder spring
4	Friction plate – 7 off	9 Spider	14 Spring seat
5	Plain plate – 6 off	10 Spring – 4 off	

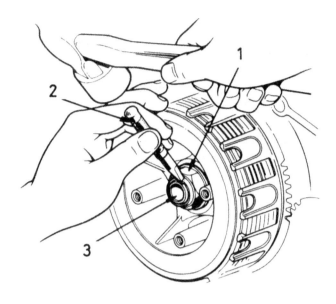

Fig. 1.7 Method of clutch centre nut retention

1 *Nut*
2 *Punch*
3 *Mainshaft*

12 Dismantling the engine/gearbox unit: separating the crankcase halves

1 Crankcase separation is necessary before attention can be given to the crankshaft, balancer shafts and drive and gearbox components. The operation can only be carried out after engine removal and preliminary dismantling as described in Sections 4 to 11 of this Chapter.
2 Working on the right-hand side of the unit, slacken and remove the balancer shaft holder lock bolt and disengage and remove the adjacent spring. Remove the tensioner anchor bolt and lift the tensioner clear of the cam chain tunnel. Displace the small oil feed pipe which runs across the centre of the tunnel to allow the cam chain to be disengaged from the crankshaft and removed. As the chain is lifted clear it should be marked clearly to indicate the outer face and the normal direction of rotation, unless a new chain is to be fitted. The incorrect fitting of a part-worn chain can result in rapid wear of both it and the sprockets which will usually become very noisy in operation.
3 On the left-hand side of the unit, remove the two bolts which secure the balancer chain guide to the crankcase, and lift the guide chain away. If necessary, turn the balancer shaft holder from the right-hand end to obtain chain free play.
4 Invert the crankcase assembly on the work bench to gain access to the lower crankcase bolts. A total of eight bolts will be found and these should be slackened evenly and progressively, and then removed. Place the crankcase assembly upright once more, and remove the eight remaining crankcase bolts.
5 Once all the sixteen bolts have been removed, the crankcase halves can be separated. Before proceeding with this stage it should be noted that the balancer chain runs between the upper and lower crankcase halves and will impede separation. It will be necessary to pause during separation to remove the chain from the rear balancer on the gearbox mainshaft, and some assistance in this would prove helpful.
6 If separation proves difficult, try tapping around the joint area with a soft-faced mallet. This will help to break the

sometimes tenacious hold of the jointing compound. If this fails, use a hammer and a hardwood block to drive the casing halves apart, taking care to select robust areas of the casing and avoiding excessive force. Once the joint begins to separate removal is quite straightforward. Lift the upper casing half a few inches clear of the lower crankcase, then lift the gearbox mainshaft and disengage the balancer chain. The upper casing can now be placed to one side.

13 Dismantling the engine/gearbox unt: removing the front balancer and gear clusters

1 With the separated crankcase halves laid out on the workbench, the various components and assemblies can be removed as follows. The upper crankcase half contains the front balancer shaft and holder. Remove the single circlip which retains the balancer sprocket to the shaft end. The sprocket and drive chain can now be slid off and placed to one side. Push the shaft and holder through from the left-hand side and remove them as a unit. The oil separator plate need not be disturbed unless it requires specific attention.
2 Moving on to the lower crankcase half, remove the crankshaft assembly, noting the half-ring which locates the right-hand main bearing. Lift out the gearbox mainshaft assembly, followed by the layshaft assembly. Place these in their correct relative positions to await further dismantling or reassembly.

14 Dismantling the engine/gearbox unit: removing the gear selector mechanism

1 The external components of the gear selector mechanism can be removed with the engine unit installed, should the need arise. In the case of the selector drum and forks, however, it will be necessary to remove the engine unit and separate the crankcase halves as described in Sections 4 to 13.
2 The external components of the gearchange mechanism include the selector shaft and return spring, the selector plate and pressure spring and the detent stopper arm. All of these components operate on the protruding end of the selector drum, the tangs on the selector plate turning the drum to the next gear position when the gear change pedal is operated, and the detent stopper holding the drum in the correct position for each gear.
3 The selector shaft forms the basis of an assembly which incorporates the return spring, selector plate and its pressure spring. The assembly is removed by pulling it clear of the casing. The stopper arm is retained by a pivot bolt and incorporates a small spring which holds it against the selector drum cam. It can be removed after the bolt has been removed and spring pressure released. Care must be taken not to damage the rather frail neutral switch contact (where fitted) which is mounted on the end of the selector drum. If necessary, the switch may be removed after releasing the central retaining bolt. No further dismantling of the selector mechanism is possible until the crankcase halves have been separated.
4 Once the crankcase halves have been separated the selector fork shafts can be displaced and the selector forks removed. It is important that the forks are refitted in their original locations, so to prevent confusion during reassembly place each one on its correct shaft and in its normal position. It is also useful to degrease and mark each fork using a spirit-based felt marker.
5 The selector drum can be removed next. It is retained at the left-hand end by a semi-circular retainer plate which is secured by two countersunk cross-head screws. These may well prove to be fairly tight and are normally coated with a thread locking compound to prevent loosening in use, so an impact driver will probably be required to avoid damage to the screw heads during removal. Once the retainer plate has been released the selector drum and bearing can be withdrawn from the casing.

12.3 Remove balancer chain guide block

14.4 Make note of fork positions during removal

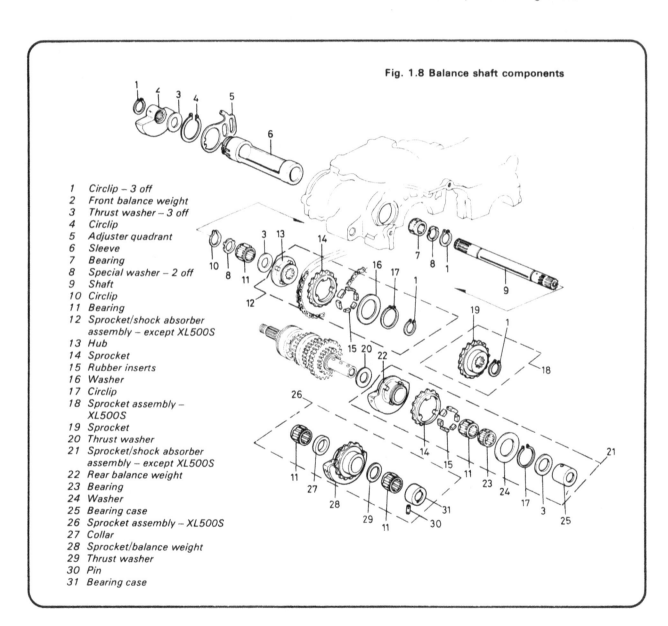

Fig. 1.8 Balance shaft components

1 Circlip – 3 off
2 Front balance weight
3 Thrust washer – 3 off
4 Circlip
5 Adjuster quadrant
6 Sleeve
7 Bearing
8 Special washer – 2 off
9 Shaft
10 Circlip
11 Bearing
12 Sprocket/shock absorber
    assembly – except XL500S
13 Hub
14 Sprocket
15 Rubber inserts
16 Washer
17 Circlip
18 Sprocket assembly –
    XL500S
19 Sprocket
20 Thrust washer
21 Sprocket/shock absorber
    assembly – except XL500S
22 Rear balance weight
23 Bearing
24 Washer
25 Bearing case
26 Sprocket assembly – XL500S
27 Collar
28 Sprocket/balance weight
29 Thrust washer
30 Pin
31 Bearing case

15 Dismantling the engine/gearbox unit: removing the kickstart shaft assembly and return spring

1 The above components can be removed only after the separation of the crankcase halves and the removal of the gearbox components as described in the preceding Sections. Start by removing the decompressor cam from the end of the kickstart shaft, together with the plain thrustwasher which precedes it. The spring and spring seat should also be slid off the shaft end.

2 Working from the inner end of the shaft, release the circlip and thrust washer from the shaft end. This will free the spring retainer which can be withdrawn from the inside of the kickstart return spring. Grasp the outer end of the spring with a strong pair of pliers, and release it from the stop. When spring tension has been released, disengage the inner tang of the spring and remove it. The shaft, pinion and ratchet can be displaced outwards and removed from the casing.

16 Examination and renovation: general

1 Before examining the parts of the dismantled engine unit for wear, it is essential that they should be cleaned thoroughly. Use a paraffin/petrol mix to remove all traces of old oil and sludge that may have accumulated within the engine.

2 Examine the crankcase castings for cracks or other signs of damage. If a crack is discovered, it will require professional repair.

3 Examine carefully each part to determine the extent of wear, if necessary checking with the tolerance figures listed in the Specifications Section of this Chapter, or accompanying the text.

4 Use a clean, lint-free rag for cleaning and drying the various components, otherwise there is a risk of small particles obstructing the oilways.

5 Should any studs or internal threads require repair, now is the appropriate time to attend to them. Where internal threads are stripped or badly worn, it is preferable to use a thread insert. The most common of these is the Helicoil type. The damaged thread is drilled oversize and then tapped to accept a diamond

15.2 Displace spring using pliers or spanner as shown

section wire thread insert. In most cases the original fastener can be used in the restored thread.

17 Examination and renovation: crankcase and fittings

1 The remaining fittings on the crankcase halves should be removed prior to cleaning and examination. This applies particularly to the half-rings and dowels, which should be removed and marked by placing them in bags with suitable labels to indicate their correct location as an aid to reassembly.

2 The crankcase halves should be thoroughly degreased, using one of the proprietary water-soluble degreasing solutions such as Gunk. When clean and dry a careful examination should be made, looking for signs of cracks or other damage. Any such fault will probably require either professional repair or renewal of the crankcases as a pair. Note that any damage around the various bearing bosses will normally indicate that renewal is

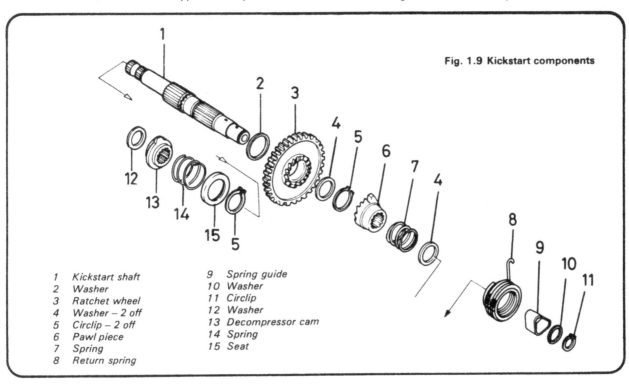
Fig. 1.9 Kickstart components

1 Kickstart shaft
2 Washer
3 Ratchet wheel
4 Washer – 2 off
5 Circlip – 2 off
6 Pawl piece
7 Spring
8 Return spring
9 Spring guide
10 Washer
11 Circlip
12 Washer
13 Decompressor cam
14 Spring
15 Seat

necessary, because a small discrepancy in these areas can result in serious mis-alignment of the shaft concerned. It is important to check crankcase condition at the earliest opportunity, because this will permit remedial action to be taken and any necessary machining or welding to be done whilst attention is turned to the remaining engine parts.

3   As mentioned previously, badly worn or damaged threads can be reclaimed by fitting a thread insert. This is a simple and inexpensive task, but one which requires the correct taps and fitting tools. It follows that the various threads should be checked and the cases taken to a local engineering works or motorcycle dealer offering this service so that repair can take place while remaining engine parts are checked.

18 Examination and renovation: crankshaft assembly

1   Check the crankshaft assembly visually for use, paying particular attention to the slot for the Woodruff key and to the threads at each end of the mainshaft. Should these have become damaged, specialist help will be needed to reclaim them.

2   The connecting rod should be checked for big-end bearing play. A small amount of end float is normal, but any up and down movement will necessitate renewal.

3   Grasp the connecting rod and pull it firmly up and down. Any movement will soon become evident. Be careful that end-float is not mistaken for wear. Should the big-nd bearing be worn it will be necessary to purchase a new replacement crankshaft assembly.

4   If measuring facilities are available, set the crankshaft in V-blocks and check the big-end bearing radial clearance using a dial gauge mounted on a suitable stand. Big-end axial clearance (end float) can be checked using feeler gauges. If either clearance exceeds the service limits given in the Specifications, it will be necessary to fit a new crankshaft. Honda do not supply new big-end bearings or connecting rods, so bearing replacement is not practicable.

5   Assuming that the big-end bearing is in good order, attention should be turned to the rest of the connecting rod. Visually check the rod for straightness, particularly if the engine is being rebuilt after a seizure or other catastrophe. Look also for signs of cracking. This is extremely unlikely, but worthwhile checking. Spotting a hairline crack at this stage may save the engine from an untimely end.

6   Check the fit of the gudgeon pin in the small-end eye. It should be a light sliding fit with no evidence of radial play. In the unlikely event that this condition is evident, the connecting rod, and thus the crankshaft assembly, will require renewal as no bush is fitted. It is recommended that the advice of a Honda Service Agent is sought, because he will have the necessary experience to advise on the best course of action.

7   Failure of the main bearings is usually evident in the form of an audible rumble from the bottom of the engine, accompanied by vibration. The vibration will be most noticeable through the footrests.

8   The crankshaft main bearings are of the journal ball type. If wear is evident in the form of play or if the bearings feel rough as they are rotated, replacement is necessary. Each bearing is positioned against the flywheel face and is a tight interference fit on the crank mainpin. Because of the small clearance between the bearing inner face and the flywheel and because of the tight fit, bearing removal using a conventional two- or three-legged puller is not a practical proposition. To add to the problem each bearing is flanked on the outside by a chain sprocket which is also a very tight fit. In view of this it is suggested that the flywheel assembly be placed in the hands of a competent engineer or motorcycle repair specialist who is equipped with a hydraulic press of the type suitable for this type of work. It should be noted that each sprocket is placed in the correct position to allow camshaft timing or balance weight timing. On reassembly ensure that the sprockets are fitted correctly. Refer to Sections 19 and 25 for further details.

19 Examination and renovation: balancer mechanism

1   The balancer mechanism fitted to the XL and XR models is of robust construction and is unlikely to require much attention during its normal service life which should be equal to that of the engine unit in general. Wear or damage may, however, result where the supply of lubricating oil has become limited. This could lead to excessive clearance due to wear in the needle roller bearings which support the rear balance weight or the similar bearing which carries the front balancer shaft. In extreme cases the out-of-balance forces inherent in the system could quickly destroy the bearings, causing them to break up. It will be appreciated that this is a far from desirable situation because the resulting debris would quickly destroy other engine components.

2   To check for wear during overhaul it is necessary to measure the internal diameter of the rear balance weight, and the internal and external diameters of the front balancer holder. These are as follows.

Front balancer holder
Internal diameter	26.007 – 26.020 mm (1.0239 – 1.0244 in)
Wear limit	26.05 mm (1.026 in)
External diameter	39.964 – 39.980 mm (1.5734 – 1.5740 in)
Wear limit	39.91 mm (1.571 in)

Rear balance weight
Internal diameter	26.007 – 26.020 mm (1.0239 – 1.0244 in)
Wear limit	25.05 mm (1.026 in)

3   The front balancer adjuster flange is retained by a large circlip to the end of the balancer holder. If it is necessary to remove it for any reason, make a note of its position in relation to the holder so that it can be refitted in the same place. In practise, it is unlikely that the flange will need to be disturbed.

4   Examine the balancer chain and sprocket teeth for signs of wear or damage. Neither condition is common unless the chain tension has been badly adjusted, in which case the chain will become loose and stretched and must be renewed. The sprockets must be renewed if the teeth are obviously hooked or chipped. On no account should worn or damaged sprockets be reused in view of the risk of rapid chain wear or breakage that might result. The chain should be renewed together with the sprockets to ensure long and reliable service.

5   The balancer drive sprocket is a tight fit on the crankshaft and can only be safely removed using a hydraulic press. Attempts at removal using a sprocket puller are likely to end in damage being done to the sprocket or crankshaft end. It is suggested that the flywheel assembly be placed in the hands of a competent engineer for this work to be carried out. Note that when refitting the sprocket to the crankshaft it must be placed in a particular position to ensure that subsequent balance weight timing is correct. The sprocket must be fitted so that the punch mark provided on the sprocket outer face is exactly in line with the centre of the keyway in the crank mainshaft. To aid assembly make a scribed line on the sprocket boss so that correct alignment can be seen easily.

6   All of the 250 models employ a rubber block and vane type shock absorber in the front and rear balancer sprockets. A similar arrangement applies to the front sprocket, and in some cases the rear sprocket of the XR500. In the case of the XL500 S plain sockets are fitted to both balancers. Where a shock absorber arrangement is used, the sprocket and hub are separate components held together by a large diameter circlip. After high mileages have been covered the rubber blocks may tend to become compressed and will allow play in the balancer system. If this is the case, remove the circlip and separate the sprocket from the hub. The fitting of new rubber blocks will be facilitated by lubricating them with petrol (gasoline) or a rubber lubricant. Note that index marks are provided on the sprocket and boss or hub, and these must align when assembled.

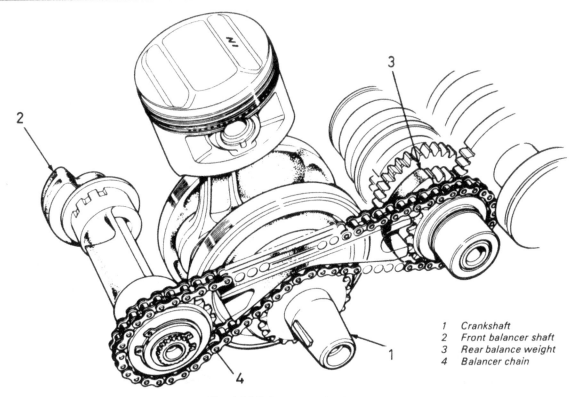

1 Crankshaft
2 Front balancer shaft
3 Rear balance weight
4 Balancer chain

Fig. 1.10 Balancer mechanism

20 Examination and renovation: cylinder barrel

1 The usual indications of a badly worn cylinder and piston are excessive oil consumption and piston slap, a metallic rattle that occurs when there is little or no load on the engine. If the top of the bore of the cylinder barrel is examined carefully, it will be found that there is a ridge on the thrust side, the depth of which will vary according to the amount of wear that has taken place. This marks the limit of travel of the uppermost piston ring.

2 If possible, an internal micrometer should be used to obtain an accurate measurement of the amount of wear which has taken place in the bore. Measurements should be made at various points within the bore, and the measurement across the most worn part of the bore compared with that of an unworn portion, e.g. the lowest part of the bore. The difference between the two readings should not exceed 0.10 mm (0.004 in). If it is found to exceed this figure, it will be necessary to have the cylinder rebored and new oversize piston fitted.

3 If, as is likely, an internal micrometer is not available, a rough check of the amount of bore wear can be made as follows. Insert the bare piston into the bore in its normal position ie, with the arrow facing the front of the cylinder. Using feeler gauges, measure the amount of clearance between the piston and bore about $\frac{1}{2}$ in from the top of the cylinder liner, and at the front of the cylinder. If this measurement exceeds the allowable clearance as stated above, it indicates that attention is required. If desired, the barrel and piston can be taken to a Honda Service Agent for verification.

4 Check the surface of the cylinder bore for score marks or any other damage that may have resulted from an earlier engine seizure or displacement of the gudgeon pin. A rebore will be necessary to remove any deep indentations, irrespective of the amount of bore wear, otherwise a compression leak will occur.

5 Check the external cooling fins are not clogged with oil or road dirt; otherwise the engine will overheat.

21 Examination and renovation: piston and piston rings

1 If a rebore is necessary, the existing piston and rings can be disregarded because thay will be replaced with their oversize equivalents as a matter of course.

2 Remove all traces of carbon from the piston crown, using a soft scraper to ensure the surface is not marked. Finish off by polishing the crown with metal polish so that carbon does not adhere so easily in the future. Never use an emery cloth.

3 Piston wear usually occurs at the skirt or lower end of the piston and takes the form of vertical streaks or score marks on the thrust side. There may also be some variation in the thickness of the skirt.

4 The piston ring grooves may also become enlarged in use, allowing the piston rings to have greater side float. If the clearance exceeds the piston ring to groove clearance wear limit figures given in the Specifications Section, the piston will require renewal, complete with rings. It should be noted that it is unusual for this type of wear to be found in an otherwise unworn engine.

5 The piston rings will tend to lose their elasticity over a period of time, the eventual result being that they will allow combustion pressure to escape into the crankcase, thus causing a noticeable drop in power. It is recommended that the free end gap of the rings be measured, and compared with the figures given in the Specifications Section. The rings should be renewed as a set if the free end gap is lower than that given under 'Wear limit'.

6 Piston ring wear is measured by removing the ring from the piston and inserting them in the cylinder bore using the crown of the piston to locate them approximately $1\frac{1}{2}$ inches from the top of the bore. Make sure thay rest square with the bore. Measure the ring end gap using feeler gauges, and compare this reading with that given in the Specifications section. Note that it is assumed that the cylinder is not in need of a rebore, as a worn bore would produce an inaccurate reading.

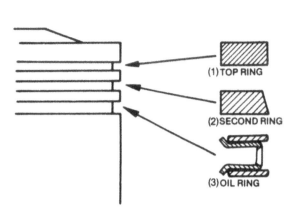

Fig. 1.11 Piston ring profiles

(1) TOP RING

(2) SECOND RING

(3) OIL RING

Fig. 1.12 Method of freeing gummed piston rings

19.1a Dismantle front balancer for examination

19.1b Note position of special tanged washers (arrowed)

19.3 Adjuster plate is secured by large circlip

22 Examination and renovation: cylinder head and valves

1   It is best to remove all carbon deposits from the combustion chambers before removing the valves from inspection and grinding in. Use a blunt end chisel or scraper so that the surfaces are not damaged. Finish off with a metal polish to achieve a smooth, shining surface. If a mirror finish is required a high speed felt mop and polishing soap may be used. A chuck attached to a flexible drive will facilitate the polishing operation.

2   A valve spring compression tool must be used to compress each set of valves in turn, thereby allowing the split collets to be removed from the valve collar and the valve springs and collars to be freed. Keep each set of parts separate so that they can be replaced in the correct location. This is best done by placing each valve, springs and spring seat in a marked plastic bag or similar to indicate its correct position. It is essential that the pairs of inlet valves and exhaust valves do not become interchanged.

3   Before attention is turned to the valves and valve seats it is important to check for wear between the valve stems and guides. The appropriate nominal and service limit figures can be found in the Specifications section of this Chapter. Measure the valve stem at the point of greatest wear and then measure

again at right-angles to the first measurement. If the valve stem is below the service limit it must be renewed. The valve stem/guide clearance can be measured with the use of a dial gauge and a new valve. Place the new valve into the guide and measure the amount of shake with the dial gauge tip resting against the top of the stem. If the amount of wear is greater that the wear limit, the guide must be renewed.

4   To remove an old valve guide, place the cylinder head in an oven and heat it to about 100°C, taking care to ensure warpage does not occur through overheating or uneven heating. The old guide can now be tapped out from the cylinder side. The correct drift should be shouldered with the smaller diameter the same size as the valve stem and the larger diameter slightly smaller than the OD of the valve guide. If a suitable drift is not available a plain brass drift may be utilised with great care. To aid removal of the valve guide and to help prevent broaching of the cylinder head material, all carbon deposits on the portion of the valve guide projecting into the port should be cleaned off prior to guide removal. Each valve guide is fitted with an oil seal to ensure perfect sealing. The oil seal must be replaced with a new components. New guides should be fitted with the head at the same heat as for removal. Note that after fitting new guides the valve seats must be re-cut using a 45° cutter or stone to ensure that each seat is correctly aligned with the guide axis.

5   Valve grinding is a simple task. Commence by smearing a trace of fine valve grinding compound (carborundum paste) on the valve seat and apply a suction tool to the head of the valve. Oil the valve stem and insert the valve in the guide so that the two surfaces to be ground in make contact with one another. With a semi-rotary motion, grind in the valve head to the seat, using a backward and forward action. Lift the valve occasionally so that the grinding compound is distributed evenly. Repeat the application until an unbroken ring of light grey matt finish is obtained on both valve and seat. This denotes the grinding operation is now complete. Before passing to the next valve, make sure that all traces of the valve grinding compound have been removed from both the valve and its seat and that none has entered the valve guide. If this precaution is not observed rapid wear will take place due to the highly abrasive nature of the carborundum base.

6   When deep pits are encountered, it will be necessary to use a valve refacing machine and a valve seat cutter, set to an angle of 45°. Never resort to excessive grinding because this will only pocket the valves in the head and lead to reduced engine efficiency. If after cutting the seat it is found that the seat width exceeds 2.0 mm, the width should be restored to the correct range of 1.2 – 1.4 mm (0.05 – 0.06 in) using first a 32°, and then a 60° cutter or stone. Because of the expense of such equipment and the expertise required for its satisfactory use it is recommended that seat restoration be placed in the hands of an expert. If there is any doubt about the condition of a valve, fit a new one.

7   Examine the condition of the valve collets and the groove on the valve stem in which they seat. If there is any sign of damage, new parts should be fitted. Check that the valve spring collar is not cracked. If the collets work loose or the collar splits whilst the engine is running, a valve could drop into the cylinder and cause extensive damage.

8   The condition of the valve springs can be assessed by measuring their free lengths and comparing the readings with those specified. If the engine is being overhauled after many miles of use, it is usually worthwhile renewing the valve springs as a matter of course.

9   Reassemble the valve and valve springs by reversing the dismantling procedure. Fit new oil seals to each valve guide and oil both the valve stem and the valve guide, prior to reassembly. Take special care to ensure the valve guide oil seal is not damaged when the valve is inserted. As a final check after assembly, give the end of each valve stem a sharp tap with a hammer, to make sure the split collets have located correctly.

10 Check the cylinder head for straightness, especially if it has shown a tendency to leak oil at the cylinder head joint. If there

is any evidence of warpage, provided it is not too great, the cylinder head must be either machined flat or a new head fitted. Most cases of cylinder head warpage can be traced to unequal tensioning of the cylinder head nuts and bolts by tightening them in incorrect sequence.

11 Mild cases of cylinder head warpage can be cured by lapping the cylinder head on a sheet of abrasive paper placed on a surface plate or sheet of plate glass. It will be necessary to remove the two projecting studs prior to this operation. Move the head with an oscillatory movement over the abrasive paper, removing no more than the minimum amount of cylinder head material required to correct the warpage. Once the mating face is flat, finish off using a very fine abrasive paper to produce a smooth matt grey finish.

Fig. 1.13 Valve seat re-cutting angles

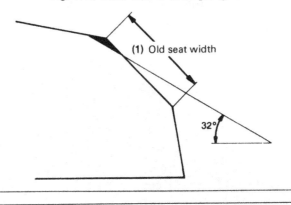

(1) Old seat width

32°

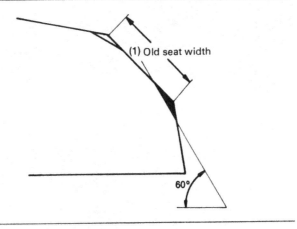

(1) Old seat width

60°

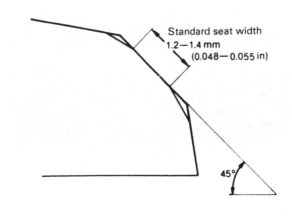

Standard seat width
1.2 – 1.4 mm
(0.048 – 0.055 in)

45°

22.2a Compress valve springs and displace collet halves

22.2b Remove spring seat and springs...

22.2c ...followed by lower spring seat

22.2d Valve can now be removed

22.9a Fit new valve stem oil seals

22.9b Fit valve springs as shown

22.9c Compress spring and fit collet halves in recess

23 Examination and renovation: rocker arms and spindles

1   The Honda XL and XR models employ two forked rocker arms to operate the four valves. The arms pivot on hardened steel spindles which are located in the cylinder head cover. Each spindle is retained by a small dowel pin which passes through its left-hand end via the cylinder head cover casting. The pins can be removed by pulling them out with a pair of pliers, and this will normally suffice to remove them. On occasions, however, it may prove helpful to heat the cover by immersing it in boiling water to assist in the removal operation. It is found that an insufficient amount of the pin protrudes from the casing to allow a good purchase to be made, removal will prove very difficult. It is recommended that the rocker cover be taken to a Honda Service Agent whose expertise can be brought to bear on the problem.

2   After the dowel pins have been removed, the spindles can be displaced and the rockers and thrust washers released. It may prove useful to tap the cylinder head screw to jar the spindles clear, using a soft-faced mallet to avoid damage to the soft alloy. Remove the rockers and spring washers and place them back on their respective shafts to avoid any part becoming interchanged.

3   Check the rocker arms for undue wear on their spindles and renew any that show excessive play. Examine each rocker arm where it bears on the cam and the opposite end which bears on the valve stem head. Arms that are badly hammered or worn should be renewed. Slight wear marks may be stoned out with an oil carborundum stone, but remember that if too much metal is removed it will not only weaken the component but may make correct valve clearance adjustment difficult. Further, the depth of hardening may be exceeded and, therefore, subsequent wear will be rapid.

24 Examination and renovation: camshaft and bearing surfaces

1   The camshaft should be examined visually for wear, which will probably be most evident on the ramps of each cam and where the cam contour changes sharply. Also check the bearing surfaces for obvious wear and scoring. Cam lift can be checked by measuring the height of the cam from the bottom of the base circle to the top of the lobe. If the measurement is less than the service limit the opening of that particular valve will be reduced resulting in poor performance.

2   Measure the diameter of each bearing journal with a micrometer or vernier gauge. If the diameter is less than the service limit, renew the camshaft.

3   The camshaft bears directly on the cylinder head material, there being no separate bearings. Check the bearing surfaces for wear and scoring. The clearance between the camshaft bearing journals and the aluminium bearing surfaces may be checked by measuring the journals and bearing surfaces, using an internal and external micrometer and calculating the amount of clearance. If the clearance is greater than given for the service limit the recommended course is to replace the camshaft. If bad scuffing is evident on the camshaft bearing surfaces of the cylinder head, due to a lubrication failure, the only remedy is to renew the cylinder head, and the camshaft if it transpires that it has been damaged also.

4   In the case of the older machines that are well out of their warranty period, it may be worth considering a conversion to needle roller camshaft bearings as an alternative to the renewal of the cylinder head, cover and camshaft. This will offer a saving over the cost of renewing the worn parts and has the advantage of being easily and inexpensively repaired should the bearings wear in the future. A number of engineering companies will undertake this type of work, and often advertise in motorcycle magazines. Note that this type of conversion is not undertaken by or condoned by Honda.

23.1 Rocker shafts are retained by dowel pins (arrowed)

25 Examination and renovation: cam chain, sprockets and tensioner

1   After high mileages have been covered it may become apparent that the cam chain tensioner is unable to compensate for wear and stretch in the cam chain, leading to noisy operation despite frequent adjustment. Once this stage is reached it is essential that the cam chain is renewed before severe wear or damage to the related components results.

2   The cam chain is of the Morse or Hy-Vo type and will normally last for a considerable time, provided that tension adjustment receives regular attention. No specific wear limits for the chain or tensioner are given, so it must be assumed that one or both require renewal if adjustment has failed to correct noisy operation. If the engine has been stripped for overhaul and the chain seems to be at or near its maximum adjustment, it is worthwhile renewing it as a precautionary measure. Note that a used chain must always be fitted in the same position as it was prior to removal, as the reversal of chain direction will cause accelerated wear and may cause noisy operation. This does not apply to new chains.

3   Wear in the tensioner blade or the cam chain guide will be self evident, and if either component is deeply grooved it should

be renewed. In extreme cases, a combination of a badly worn tensioner blade and guide will prevent adjustment of a serviceable chain. On no account should the pad material be allowed to wear through to the backing metal.

4   The camshaft mounted chain sprocket is bolted to the camshaft and in consequence is easily renewable if the teeth become hooked, worn, chipped or broken. The lower sprocket is a tight interference fit on the crankshaft and can only be safely removed using a hydraulic press. Attempts at removal using a sprocket puller are likely to end in damage to the sprocket or crankshaft end. It is recommended that the flywheel assembly be placed in the hands of a competent engineer for this work to be carried out. When fitting a new sprocket it must be positioned correctly so that cam timing can be maintained correctly. The sprocket must be so placed that the centre line through any one tooth is in exact alignment with the centre line of the drive pin in the crank mainshaft. To aid assembly scribe a line on the sprocket boss so that the alignment can be checked visually with ease.

5   If the sprocket(s) are renewed, the chain should be renewed at the same time. It is bad practise to run old and new parts together since the rate of wear will be accelerated.

26 Examination and renovation: clutch assembly

1   After an extended period of service the clutch linings will wear and promote clutch slip. The limit of wear measured across each inserted and the standard measurement is as follows:

	Standard	Service limit
Clutch plate thickness	2.62-2.78 mm (0.102.0.109 in)	2.3 mm (0.09 in)

When the overall width reaches the limit, the inserted plates must be renewed, preferably as a complete set.

2   The plain plates should not show any excess heating (blueing). Check the warpage of each plate glass or surface plate and a feeler guage. The maximum allowable warpage is 0.30 mm (0.012 in).

3   The clutch springs will lose tension after a period of use, and should be renewed as a precaution if clutch slip has been evident and the friction plates are within limits. The free length of the clutch springs give a good indication of condition, and this should be checked and compared with the figures given in the Specifications Section.

4   Examine the clutch assembly for burrs or indentations on the edges of the protruding tongues of the inserted plates

and/or slots worn in the edges of the outer drum with which they engage. Similar wear can occur between the inner tongues of the plain clutch plates and the slots in the clutch inner drum. Wear of this nature will cause clutch drag and slow disengagement during gear changes, since the parts will become trapped and will not free fully when the clutch is withdrawn. A small amount of wear can be corrected by dressing with a fine file; more extensive wear will necessitate renewal of the worn parts.

5   The clutch release mechanism takes the form of a spindle running in the right-hand outer casing, the shaped end of which bears on the clutch release pushrod when the handlebar lever is operated. The mechanism is of robust construction and requires no attention during normal maintenance or overhauls.

27 Examination and renovation: gearbox components

1   Examine each of the gear pinions to ensure that there are no chipped or broken teeth and that the dogs on the end of the pinions are not rounded. Gear pinions with any of these defects must be renewed; there is no satisfactory method of reclaiming them.

2   The gearbox bearings must be free from play and show no signs of roughness when they are rotated. The bearings should be washed in petrol and then dried. Also check for pitting on the roller tracks.

3   It is advisable to renew the gearbox oil seals irrespective of their condition. Should a re-used seal fail at a later date, a considerable amount of work is involved to gain access to renew it.

4   Check the gear selector rods for straightness by rolling them on a sheet of plate glass. A bent rod will cause difficulty in selecting gears and will make the gear change particularly heavy.

5   The selector forks should be examined closely, to ensure that they are not bent or badly worn. Under normal conditions, the gear selector mechanism is unlikely to wear quickly, unless the gearbox oil level has been allowed to become low.

6   The tracks in the selector drum, with which the selector forks engage, should not show any undue signs of wear unless neglect has led to under lubrication of the gearbox. Check the tension of the gearchange pawl, gearchange arm and drum stopper arm springs. Weakness in the springs will lead to imprecise gear selection. Check the condition of the gear stopper roller and pins in the change drum end with which it engages. It is unlikely that wear will take place here except after considerable mileage.

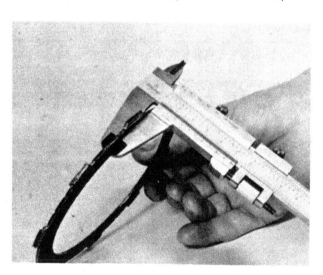

26.1 Measure thickness of clutch friction plates

26.5 Clutch release mechanism is unlikely to require attention

7   Check the condition of the kickstart components. If slipping has been encountered a worn ratchet and pawl will invariable be traced as the cause. Any other damage or wear to the components will be self-evident. If either the ratchet or pawl is found to be faulty, both components must be replaced as a pair. Examine the kickstart return spring which should be renewed if there is any doubt about its condition.

8   If it proves necessary to dismantle the gearbox shafts for further examination or to renew worn or damaged parts, reference should be made to Fig. 1.14 and the photographic sequence which shows the correct assembly sequence (see Section 29). When rebuilding the shafts it is advisable to use new thrust washers and circlips throughout, and these should be obtained when purchasing the new seals and gaskets required for reassembly.

28 Engine reassembly: general

1   Before reassembly of the engine/gear unit is commenced, the various component parts should be cleaned thoroughly and placed on a sheet of clean paper, close to the working area.

2   Make sure all traces of old gaskets have been removed and that the mating surfaces are clean and undamaged. One of the best ways to remove old gasket cement is to apply a rag soaked in methylated spirit. This acts as a solvent and will ensure that the cement is removed without resort to scraping and the consequent risk of damage.

3   Gather together all of the necessary tools and have available an oil can fitted with clean engine oil. Make sure all new gaskets and oil seals are to hand, also all replacement parts required. Nothing is more frustrating than having to stop in the middle of a reassembly sequence because a vital gasket or replacement has been overlooked.

4   Make sure that the reassembly area is clean and that there is adequate working space. Refer to the torque and clearance settings wherever they are given. Many of the smaller bolts are easily sheared if over-tightened. Always use the correct size screwdriver bit for the crosshead screws and never an ordinary screwdriver or punch. If the existing screws show evidence of maltreatment in the past, it is advisable to renew them as a complete set.

27.1 Check condition of all gear teeth and bearings

27.5 Measure selector fork widths to assess wear

27.6 Check condition of selector shaft assembly

27.7 Check kickstart ratchet and gear teeth

29 Engine reassembly: rebuilding the gearbox clusters

1   If the gearbox components have been dismantled for examination and renewal, it is essential that they are rebuilt in the correct order to ensure proper operation of the gearbox. Use the accompanying photographic sequence and line drawings as aids to the identification of components and their correct relative positions.

Layshaft (output shaft) assembly

2   Take the layshaft complete with its left-hand bearing and 2nd gear pinion, and fit the 5th gear pinion with its selector fork groove facing the right-hand end of the shaft.
3   Slide the 3rd gear pinion into position, followed by the splined thrust washer. Retain them with a circlip.
4   The 4th gear pinion is fitted next, with its selector groove facing inwards (towards the left), followed by a plain thrust washer.
5   Fit the large 1st gear pinion, ensuring that it engages the dogs of the preceding gear.
6   Place the right-hand layshaft bearing (needle roller type) over the shaft end.
7   Fit a new oil seal to the left-hand end of the shaft.

Mainshaft (input shaft) assembly

8   Where necessary, fit a new bearing to the right-hand (clutch) end of the mainshaft, ensuring that it butts squarely against the integral 1st gear pinion. Note that the locating groove must face outwards.
9   Place the 4th gear pinion in position, ensuring that the dogs face towards the left-hand end of the shaft. Place a splined thrust washer against the 4th gear pinion and secure it with a circlip.
10   The 3rd gear pinion should be fitted next, with its selector groove facing inwards, or towards the right-hand end of the shaft.
11   This is followed by a circlip and a thrust washer, and then the 5th gear pinion. Fit another thrustwasher and circlip to retain it.
12   Slide the small 2nd gear pinion into place, followed by the large plain thrustwasher which locates it.
13   Assembly is completed by fitting the bearings and the balance weight. The bearing arrangement varies between models. In all cases, a wide needle roller bearing is fitted first. This is followed by either a narrow needle roller bearing or a spacer of similar width. Fit the balancer weight with the sprocket facing outwards, then finish the assembly by fitting the outer needle roller bearing preceded by the remaining large thrust washer.

29.2a The layshaft, complete with LH bearing and 2nd gear

29.2b Fit 5th gear pinion with selector groove outwards

29.3 Fit 3rd gear pinion, thrust washer and circlip

29.4a Fit 4th gear pinion with selector groove inwards...

29.4b ...followed by the plain thrustwasher as shown

29.5 Fit 1st gear bush, followed by pinion

29.6 Fit the needle roller bearing assembly

29.7 Place a new seal on the left-hand shaft end

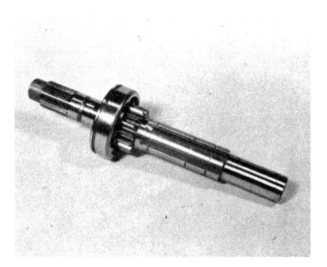
29.8 Mainshaft with RH bearing in position

29.9a Fit the 4th gear pinion with dogs facing outwards

29.9b Fit splined thrustwasher and secure with circlip

29.10 Fit 3rd gear pinion with selector groove inwards

29.11a Locate circlip and fit thrustwasher...

29.11b ...followed by 5th gear pinion

29.11c Retain the pinion with a thrustwasher and circlip

29.12a Slide the 2nd gear pinion into position...

29.12b ...followed by large plain thrustwasher

29.13a Fit needle roller bearing, then narrow bearing or spacer

29.13b Place balancer weight in position...

29.13c ...followed by thrustwasher and bearing

30 Engine reassembly: refitting the half-rings, dowels and kickstart mechanism

1   Place the lower crankcase half on the workbench and fit the locating half-rings to the grooves provided in the left-hand layshaft boss and the right-hand mainshaft and crankshaft bosses. Place the locating dowels in the appropriate holes at the front and rear of the jointing surface.

2   Refit the kickstart stopper plate to the outer face of the lower crankcase half and return the spring anchor pin. The pin is secured by a nut on the inside of the casing. The anchor pin and the retaining bolt should be tightened to the following torque figures.

Anchor pin	2.2 – 2.8 kgf m (16 – 20 lbf ft)
Anchor pin locknut	1.6 – 2.0 kgf m (12 – 15 lbf ft)
Retaining bolt:	
X5500S	1.8 – 2.5 kgf m (13 – 18 lbf ft)
All others	0.8 – 1.2 kgf m (6 – 9 lbf ft)

3   Fit the larger circlip, thrust washer, starter pinion, thrust washer and the smaller circlip to the kickstart shaft, securing the pinion to the shaft. Place the starter ratchet over the shaft

splines, ensuring that the alignment marks on the shaft and the ratchet are correctly positioned. If this is not ensured correct kickstarter operation will not be possible. Fit the ratchet spring and thrustwasher.

4   Fit the shaft assembly into the casing, ensuring that the spring hole at the inner end faces upwards and that the oil hole is visible in its casing recess. Before proceeding further, fill the oil hole with clean engine oil. Place the return spring over the shaft and engage its inner tang in the cross-drilling in the shaft. Once engaged, fit the retaining collar between the spring and the shaft.

5   Using a stout pair of pliers, turn the free end of the spring until it can be hooked over the anchor pin. Fit the plain washer against the retaining collar and secure it with a new circlip, ensuring that its chamfered edge faces outwards. It is advisable to renew this circlip as a precaution in view of the considerable dismantling work that would be required should it become displaced in use. Check that it seats correctly.

6   Moving to the external part of the kickstart shaft assembly, place the spring seat and spring over the protruding end of the shaft. Fit the decompressor cam, ensuring that the punch marks on the cam and shaft are correctly aligned.

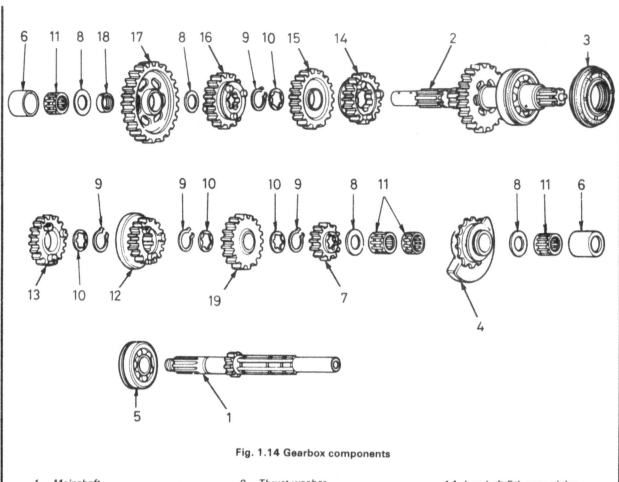

Fig. 1.14 Gearbox components

1	Mainshaft	8	Thrust washer
2	Layshaft	9	Circlip – 4 off
3	Oil seal	10	Splined washer – 4 off
4	Balance weight	11	Needle roller bearing
5	Bearing	12	Mainshaft 3rd gear pinion
6	Needle roller bearing shell	13	Mainshaft 4th gear pinion
7	Mainshaft 2nd gear pinion		

14	Layshaft 5th gear pinion
15	Layshaft 3rd gear pinion
16	Layshaft 4th gear pinion
17	Layshaft 1st gear pinion
18	Bush
19	Mainshaft 5th gear pinion

30.3a Kickstart pinion is retained by washer and circlip (arrowed)

30.3b Check that ratchet block is aligned as shown

30.4a Fit washer and spring and insert kickstart shaft

30.4b Fill oil hole (arrowed) and install return spring

30.4c Fit retaining collar as shown...

30.4d ...and retain with washer and circlip

30.5 Hook spring end over anchor bolt

30.6a Fit spring and seat over outside end of shaft

30.6b Slide cam into position, ensuring dots align (arrowed)

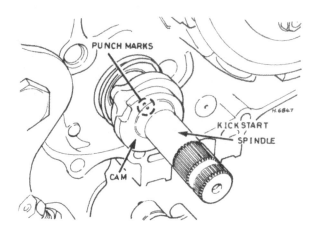

Fig. 1.15 Decompressor cam alignment marks

31 Engine reassembly: fitting the gear selector mechanism

1   Fit and lubricate the selector drum bearing, and lubricate the plain end of the selector drum. The drum can now be slid into the casing ensuring that the bearing locates correctly. Fit the bearing retainer plate, using a thread locking compound on the two countersunk screws which secure it.
2   Slide each end of the selector fork shafts into the casing, positioning the forks on the shaft as they enter the casing bores. Make sure that the forks are arranged as shown in the accompanying photographs and line drawings.
3   Place the stopper plate over the projecting end of the selector drum, noting the small dowel pin which locates the stopper plate in relation to the drum. Assemble the spacer, neutral switch contact (where fitted) and retaining bolt and washer. Note that it is possible to fit the neutral switch in two positions, and care should be taken to ensure that it is fitted correctly. The outer face of the stopper plate has four raised

dogs with which the selector claws engage. The remaining lobe of the plate is plain, and the neutral contact end should coincide with the first dog clockwise from the plain lobe. This applies irrespective of whether the straight or helical contact blade is fitted.
4   Turn the selector drum until the plain lobe of the stopper plate is almost vertical, then assemble the detent stopper arm and spring, securing it with its pivot bolt. The roller on the end of the stopper arm should engage the small neutral recess in the plate. Check that the selector shaft is correctly assembled, then install it in the casing, taking care not to damage the neutral contact during installation.
5   Temporarily refit the gearchange pedal and check that the mechanism selects each gear correctly. It will be necessary to hold each of the forks in engagement with the selector drum tracks, and these should be lubricated with engine oil to ensure free movement. When the selection has been checked, set the drum back to the neutral position.

31.1a Fit bearing to selector drum end...

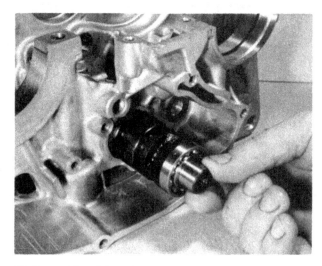

31.1b ...and install in casing

31.1c Fit retaining plate using locking fluid on screws

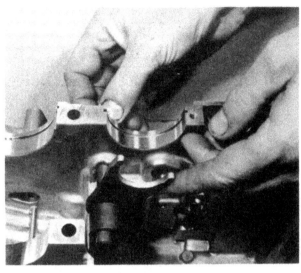

31.2a Fit selector forks and support shafts

31.2b Note positions of selector forks

31.3a Pin locates stopper plate on drum end

31.3b Fit neutral switch blade (see text)

31.4a Stopper arm is retained by shouldered bolt

31.4b Note position of circlips and thrust washers

31.5 Install selector shaft assembly as shown

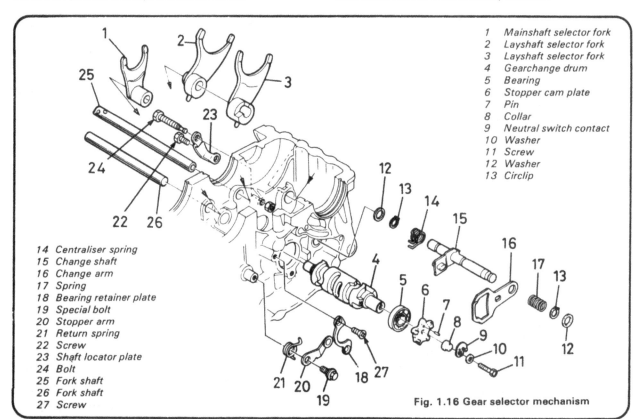

1	Mainshaft selector fork
2	Layshaft selector fork
3	Layshaft selector fork
4	Gearchange drum
5	Bearing
6	Stopper cam plate
7	Pin
8	Collar
9	Neutral switch contact
10	Washer
11	Screw
12	Washer
13	Circlip

14 Centraliser spring
15 Change shaft
16 Change arm
17 Spring
18 Bearing retainer plate
19 Special bolt
20 Stopper arm
21 Return spring
22 Screw
23 Shaft locator plate
24 Bolt
25 Fork shaft
26 Fork shaft
27 Screw

Fig. 1.16 Gear selector mechanism

32 Engine reassembly: refitting the gearbox clusters and crankshaft

1　Place the gearbox layshaft (output shaft) cluster in position, ensuring that the locating lip on the oil seal engages with its groove and that the half-ring locates in the bearing groove. The small needle-roller bearing on the right-hand end of the shaft should engage with the small locating dowel at the bottom of the casing recess, and to assist in location a pair of scribed lines on the end of the outer race should align with the gasket face.
2　Place the balancer chain around the rear balance weight sprocket, ensuring that the white-painted timing links face outwards. If the marks are indistinct or missing, lay the chain on the workbench and stretch it into an elongated loop. Mark a link at the left-hand end of the lower run, then, working from left to right, mark the 9th and 27th links in a similar fashion. Note that for timing purposes, the 1st link should coincide with the front balancer mark, the 9th link with the crankshaft and the 27th link with the rear balancer mark. The chain should therefore be arranged so that the 27th link coincides with the timing dot on the rear balancer. The gearbox mainshaft (input shaft) should now be fitted, bearing in mind the above.
3　Lower the crankshaft assembly into position in the crankcase, making sure that the bearings seat squarely and that the half-ring locates in the right-hand main bearing.

32.1a Lower layshaft assembly into casing

32.1b Check that scribe lines on bearing align as shown

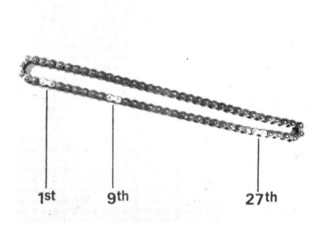

1st   9th   27th

32.2a Balancer chain should be marked for timing

32.2b 27th link should align with rear balancer dot

32.2c Lower mainshaft assembly into casing

32.3 Fit crankshaft, ensuring that half-ring engages

33 Engine reassembly: refitting the front balancer assembly and setting the balancer timing

1   If the front balancer sprocket has been dismantled for examination of the damper rubbers, reassemble it ensuring that the alignment mark on the sprocket base coincides with that of the sprocket. Ensure that the sprocket faces in the correct direction, then fit the side plate and retaining circlip.

2   Fit the circlips to each of the two inner circlip grooves on the balancer shaft, making sure that the chamfered edge of each faces inwards. Place the special tanged washer against the two circlips, then fit and lubricate the needle roller bearings. Slide the balancer weight over the splined shaft end, ensuring that the alignment mark on each coincides. Retain the weight with its circlip noting that the chamfered edge must face inwards.

3   Check that the balancer holder flange is securely retained, then slide the holder into the casing. Lubricate the balancer shaft bearings and slide the shaft assembly into the holder bore.

4   Place the upper crankcase half loosely on top of the assembled lower half. Place the front balancer sprocket in the loop of the chain extending from the rear sprocket, ensuring that the timing mark aligns with the paint-marked 1st link. Fit the thrust washer to the balancer shaft end, then align the marks on the sprocket and shaft and slide the sprocket into place. The sprocket can be secured by its circlip.

5   The crankcase halves should now be assembled 'dry' to check the balancer timing. Set the crankshaft at TDC by pulling the connecting rod fully upwards. Carefully lower the upper crankcase half checking the front and rear balancer sprockets remain in alignment with the marked links on the chain. The remaining paint-marked link (9th) should align with the dot on the crankshaft sprocket. The balancer timing is correct when, at TDC, the paint-marked links of the balancer chain correspond with their respective timing marks and the front and rear balance weights are arranged so that their flat faces align with the arrows cast into the adjacent crankcase areas. If all is correct in this respect, the timing is set accurately and the crankcase halves can be joined.

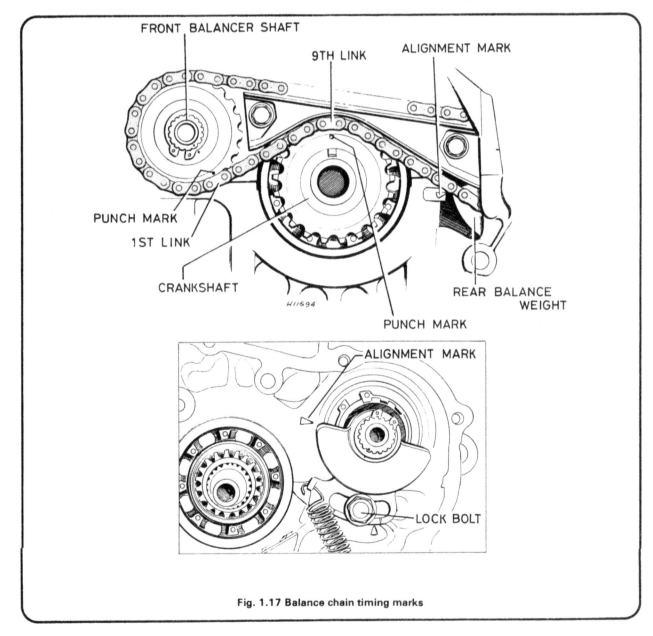

Fig. 1.17 Balance chain timing marks

33.1 Front balancer marks should align as shown

33.2 Check marks on front balancer weight (arrowed)

33.3 The front balancer and holder in position

33.4a Check alignment of dots and marked links (arrowed)

34 Engine reassembly: joining the crankcase halves

1   Having set up and checked the balancer timing as described in the preceding Section, the crankcase halves can be joined. The upper casing should be lifted clear and held or blocked in position while the mating surface of the lower crankcase is coated with a jointing compound. If essential, the balancer chain can be disengaged and the upper casing placed to one side, but do not forget that the timing will have to be re-checked if this course of action is chosen.

2   Apply a thin even film of jointing compound to the jointing face, using one of the silicone rubber based sealants. Take care not to get the compound in or near to any of the small oil passages. Leave the compound for a few minutes, then lower the upper casing half into position, ensuring that the balancer timing is preserved. Check that the two casing halves seat squarely, then fit the eight upper retaining bolts. These should be tightened in two stages to the torque setting specified below, following the tightening sequence shown in the accompanying diagrams.

33.4b Sprocket is retained by a large circlip

Upper crankcase bolt torque settings
*   6 mm bolts   1.0 – 1.4 kgf m (7 – 10 lbf ft)*
*   8 mm bolts   2.2 – 2.8 kgf m (16 – 20 lbf ft)*

3   Turn the unit over and fit the remaining crankcase bolts in a similar manner, tightening each one in two stages, following the accompanying sequence. The appropriate torque figures are as follows.

Lower crankcase bolt torque settings
6 mm bolts	*1.0 – 1.4 kgf m (7 – 10 lbf ft)*
8 mm bolts	*2.2 – 2.8 kgf m (16 – 20 lbf ft)*
9 mm bolts	*2.7 – 3.3 kgf m (20 – 24 lbf ft)*
10 mm bolts	*3.2 – 3.8 kgf m (23 – 28 lbf ft)*

4   It should be noted that the upper crankcase bolts on all models are a mixture of 6 mm and 8 mm, and these sizes are found on the lower crankcase of 250 models. In the case of the 500 models, 6 mm, 9 mm and 10 mm bolts are used on the lower crankcase. Before moving on, refit the balancer chain guide block above the crankshaft on the left-hand side of the unit. The balancer chain tension can now be set as described in Section 35.

34.3 Fit the balancer chain guide as shown

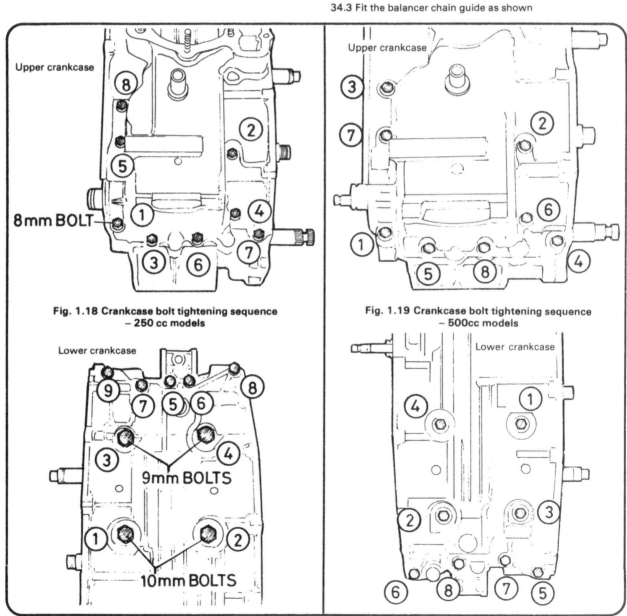

Fig. 1.18 Crankcase bolt tightening sequence
– 250 cc models

Fig. 1.19 Crankcase bolt tightening sequence
– 500cc models

35 Engine reassembly: setting the balancer chain tensioner

1   Once the crankshaft halves have been assembled, the balancer chain tension should be set to provide the correct operating conditions for the chain and sprockets. To achieve this, it is necessary to vary the distance between the front and rear balancers, and it is for this purpose that the front balancer holder is provided. The front balancer shaft runs in the eccentric bore of a tubular holder. The right-hand end of the holder terminates in an adjuster plate which is located by a single retaining bolt. When adjustment is necessary the holder is rotated in the crankcase, effectively moving the axis of the front balancer shaft forwards or backwards.

2   To adjust the chain tension it will first be necessary to fit the adjuster spring to the adjuster plate or flange. Check that the holder moves freely in the casing and turns fully anti-clockwise under spring pressure. Check that the chain is tensioned in this position.

3   The lower edge of the adjuster plate is marked by a series of lines. The balancer chain tension is set by moving the adjuster plate back (clockwise) by one graduation from the fully tensioned position. Holding this position, refit and tighten the locking bolt to 16 – 20 lbf ft (2.2 – 2.8 kgf m).

4   Occasionally it may prove impossible to obtain sufficient adjustment within the range of movement provided by the elongated slot. If this proves to be the case it will be necessary to move the adjuster plate in relation to the holder. Disconnect the adjuster spring. Release the circlip which retains the balance weight, then slide the latter off the end of the balancer shaft. Remove the plain washer which is fitted behind the balancer weight, then release the large circlip which secures the adjuster plate to the holder. It will be seen that the adjuster plate is located by tangs which engage in corresponding slots in the holder. Withdraw the adjuster plate and reposition it one slot further round (clockwise) to bring the range of adjustment within the scope of the plate.

5   Reassemble the balancer components in the reverse of the dismantling sequence, noting that the balancer timing mark must align with its counterpart on the balancer shaft. The tensioning operation can now be completed as described above.

36 Engine reassembly: refitting the oil pump

1   The oil pump may be fitted at any stage of reassembly after the selector forks have been fitted, but prior to the installation of the clutch. If the pump was dismantled for overhaul ensure that it has been assembled correctly and that the alignment dowel is in position. The pump should be primed by introducing oil into the inlet orifice whilst the pump pinion is rotated.

2   Check that the casing is perfectly clean, then fit new O-rings to the pump inlet and outlet orifices. Lower the pump into position over the projecting end of the selector fork shaft making sure that it seats squarely on the O-rings in the casing recess. Place the oil pump idler pinion in position over the selector fork shaft end, then offer up the retainer plate with the two retaining bolts. It is essential that the plate engages in the recess in the end of the selector fork shaft because it serves to locate it in the correct position to ensure an adequate oil supply to the gearbox. When everything is in position the two retaining bolts can be tightened down evenly. Check that the single cross-head screw is secured.

37 Engine reassembly: installing the clutch assembly

1   The clutch fitted to the XL and XR 250 and 500 models is essentially the same throughout the range other than the anti-judder spring and seat incorporated in the case of the larger models. The clutch plates should be built up around the clutch centre. With the larger models, fit the flat anti-judder spring seat followed by the spring itself. The outer edge of the dished spring should contact the spring seat. A special friction plate is fitted next. This has a larger internal diameter than the remaining plates and fits outside the anti-judder spring. A plain plate follows next, then a friction plate and so on fitting alternate plain and friction plates until the assembly is complete.

2   In the case of the 250 machines, the procedure is simpler, and assembly starts and finishes with a friction plate, alternating with plain plates as the assembly is built up. In either case, new plates should be coated with engine oil prior to reassembly and care should be taken to align the friction plate tangs. Finally, fit the clutch pressure plate, passing the four projecting pillars through the corresponding holes in the clutch centre.

3   Place the heavy thrust bearing onto the clutch shaft (mainshaft) and lubricate and fit the clutch bush. The assembled clutch plates can now be fitted into the clutch outer drum either before the drum is fitted to the mainshaft or, as shown in the photographic sequence, after fitting the drum. In either case the clutch plate tangs will probably have to be realigned to enter the clutch drum slots. Place the dished clutch lock washer in position, noting that the word 'OUTSIDE' should face outward. Lock the clutch in the same way as was used during dismantling, then fit and tighten the central retaining nut. Tighten the nut to the specified torque setting of 4.5 – 6.0 kgf m (33 – 43 lbf ft). Note that the 500 models employ a special locking nut which should be renewed each time the clutch is dismantled. On these models the collar on the outside of the nut should be staked into the groove in the shaft end using a hammer and a punch or chisel. See Fig. 1.7. If a natural antipathy to this barbaric treatment cannot be overcome, a thread locking compound may be used instead.

4   Fit the clutch springs, followed by the outer pressure plate, or lifter plate. Fit and tighten the bolts in a diagonal sequence so that the plate is pulled down evenly and squarely. Fit the clutch release bearing and pushrod.

5   The engine oil filter screen can now be fitted into its slot in the crankcase. Using a new gasket install the outer casing, noting that the decompressor lever must be lifted upwards as the casing is installed to allow its arm on the inside of the casing to engage the track on the kickstart's decompressor cam. When the casing is in place install the bolts and tighten them securely. Install the kickstart lever.

38 Engine reassembly: refitting the cam chain, tensioner, primary drive pinion and ignition rotor

1   Place the cam chain around the crankshaft sprocket, using the mark made during renewal as reference to ensure that the chain faces in its original direction of rotation. As mentioned previously, a part worn chain will cause rapid wear of the sprockets and excessive noise if refitted incorrectly. Needless to say, this does not apply if a new chain is used. The rest of the chain should be fed through the cam chain tunnel and rested against the crankcase. Fit the small oil feed pipe which runs across the centre of the tunnel.

2   Pass the cam chain tensioner down through the tunnel, and secure it with its pivot bolt at the lower end. Note the headed sleeve which passes through the tensioner eye and provides a bearing surface for the pivot bolt. The flanged end of the sleeve must face the crankcase.

3   The primary drive pinion should be fitted next. Note that the gear teeth are heavily chamfered on one side only, and these should face the crankcase when the pinion is positioned. Ensure that the cutaway section of the splines coincides with the small dowel pin in the crankshaft end.

4   Slide the ignition pickup rotor over the crankshaft end, turning it until the small notch in the rotor engages with the dowel pin in the crankshaft. If a new rotor or stator assembly have been fitted, check carefully to ensure that each carries a matching identification mark. If the two are mis-matched, ignition performance will suffer. Fit the oil feed pad (quill) and spring into the crankshaft end, retaining them with the small securing pin. Fit the washer and securing nut, tightening the latter to 4.5 – 6.0 kgf m (33 – 43 lbf ft). Check that the rotor moves freely against spring pressure, returning fully when released.

35.3 Set chain tension and secure lock bolt

36.2a Fit new O-rings to pump ports in casing (arrowed)

36.2b Place pump in casing recess...

36.2c ...and fit idler gear to shaft end

36.2d Retainer and shaft cut-out must coincide

37.2a Fit plates to clutch centre as shown

37.2b Assemble clutch centre, plates and pressure plate

37.3a Fit thrustwasher and clutch drum bearing

37.3b Slide clutch drum into position

37.3c Place centre assembly into drum

37.3d Dished washer should be fitted as marked

37.3e Lock crankshaft, fit primary gear and torque clutch nut

37.4 Assemble outer pressure plate and springs, followed by pushrod and bearing

38.1a Loop the cam chain around crankshaft sprocket

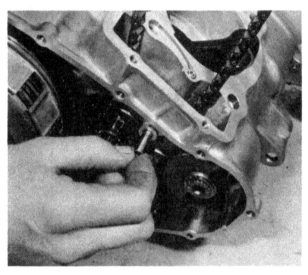

38.1b Slide oil feed pipe across centre of aperture

38.2 Assemble the chain tensioner as shown

38.3a Heavily chamfered teeth must face crankcase (arrowed)

38.3b Fit primary drive pinion over crankshaft splines

38.4a Slide ATU into place on crankshaft end

38.4b Fit oil feed quill and secure with pin

38.4c Lock crankshaft and tighten crankshaft nut

39.1 Refit alternator rotor and secure with bolt

39 Engine reassembly: refitting the alternator and left-hand outer cover

1   Fit the Woodruff key into its slot in the crankshaft end and ensure that it seats fully and squarely. Slide the alternator rotor into position. Using the same method that was employed during dismantling, hold the crankshaft to prevent its rotation, then fit and tighten the retaining bolt to 9.5 – 10.5 kgf m (69 – 76 lbf ft) XL250S and XR 250, or 10.0 – 12.0 kgf m (72 – 87 lbf ft) in the case of the XL500S and XR500 models.
2   If the stator coils were removed for any reason, they should be fitted as shown in figure 6.4. Note that the retaining plates are marked F and R denoting front and rear respectively. Check that the gasket face of the cover and crankcase are clean and dry, then fit a new gasket. Note that a locating dowel is fitted near the front of the casing and a second dowel is fitted adjacent to the gearchange shaft. Offer up the cover, fitting and tightening the hexagon-headed screws to 0.8 – 1.2 kgf m (6 – 9 lbf ft). The two inspection caps in the cover should be left at this stage to facilitate crankshaft rotation and timing checks.

40 Engine reassembly: refitting the piston, cylinder barrel and cylinder head

1   The above components can be fitted to the 250 models with the engine in or out of the frame, whilst in the case of the larger machines, it is necessary to fit the complete assembly before the unit is fitted in the frame. Commence reassembly by cleaning the mating faces of the cylinder head, barrel and crankcase to remove any residual dirt, oil or pieces of gasket. Check that the valves are installed correctly and that the necessary gaskets and O-rings are to hand before proceeding further.
2   The piston rings should be fitted to the piston, and it is important that the correct approach is adopted to avoid breakage of the very brittle rings. Starting with the lower (oil) ring, install the convoluted expander section, ensuring that the ends butt together and do not overlap or become tangled. The two steel rails, which fit either side of the expander, are fairly flexible, and do not snap easily. They should be worked into position, taking care to position the end of each rail about 30° from the ends of the expander.
3   The two compression rings differ in section, and each must be fitted in its appropriate groove. The 2nd or middle ring has a tapered working face. This must face upwards, as should the letter 'N' which is etched on the top surface of each ring, adjacent to the ring end gap. The top or 1st ring can be

identified by its plain section.

4   With a certain amount of practice, the ring gap can be spread sufficiently for the ring to be slipped into position. A safer method is to use three thin tin strips, spaced around the piston. These will allow the ring to be slid into position without fear of breakage. Arrange the ring gaps about 180° apart, and oil the rings and piston liberally.

5   Offer up the piston to the connecting rod ensuring that the IN mark on the piston crown faces rearwards. The gudgeon pin should push through fairly easily, but if it should prove stubborn, a rag soaked in hot water can be wrapped round the piston. The resulting expansion will allow the pin to be pushed through with ease. Retain the gudgeon pin with two new circlips. Resist the temptation to re-use old circlips, as they are invariably weakened during removal, and can become displaced during use, causing extensive (and expensive) engine damage. Note that the crankcase mouth should be packed with rag whilst the piston is installed, to prevent the ingress of debris or a displaced circlip.

6   Check that the cylinder barrel is clean, and that the oilways are clear. Use compressed air to blow the oilways out, or failing this, pipe cleaners or a strip of lint-free rag. Fit the two dowel pins which locate the cylinder barrel, noting that they fit around the front right-hand and rear left-hand studs. Fit a new cylinder base gasket to the crankcase, and a new O-ring around the projecting cylinder barrel liner.

7   Turn the crankshaft until the piston is at TDC, supporting the loop of the camshaft chain to prevent it bunching against the casing. Coat the piston and cylinder bore with oil to aid assembly. Fitting the barrel unaided requires a certain amount of skill, and it may be helpful to have an assistant who can lower the barrel whilst the piston and rings are guided into place. The lead in or chamfer around the bottom of the cylinder liner makes assembly much easier. Lower the barrel slowly, compressing each ring in turn as it enters the bore. It should be noted that a piston ring compressor can be utilised, and this can often prove invaluable if the job is being done unaided. When the cylinder barrel spigot has engaged all the rings remove the rag padding from the crankcase mouth.

8   As the cylinder barrel is lowered into position, pull the camshaft chain up through the tunnel. Check that the cylinder barrel seats squarely, then fit the two nuts and washers at the front and rear of the barrel and, on 500 models, the two bolts on the cam chain tunnel side. In the case of the smaller machines, the two nuts are not fitted, the barrel being retained by the two bolts only at this stage. Do not tighten any of the fasteners at this stage.

9   Pass a length of wire through the cam chain and secure it by passing it over the projecting tensioner. Lower the cam chain guide into the tunnel, ensuring that it engages correctly in the casing recess. Fit the tensioner locknut by a few turns only. Grasp the top of the tensioner and pull it fully upwards to give the maximum free play in the chain. Holding this position, secure the tensioner by tightening the locknut.

10   Fit the three cylinder head location dowels in their approximately sized holes, then fit a new cylinder head gasket. Lower the cylinder head into position, guiding the chain and tensioner through the aperture on the right-hand side. Fit the cylinder head bolts, tightening them evenly and progressively in a diagonal pattern to the following torque setting.

Cylinder head nut torque setting
XL250 S, XR250	3.5 – 4.0 kgf m (25 – 29 lbf ft)
XL500 S, XR500	2.2 – 2.8 kgf m (16 – 20 lbf ft)

Tighten the cylinder barrel nuts and/or bolts to the torque setting shown below, then recheck the cylinder head nut torque settings.

Cylinder base nut/bolt torque settings
XL250 S, XR250	
Bolts only	1.0 – 1.4 kgf m (7 – 10 lbf ft)
XL500 S, XR500	
Bolts	1.0 – 1.4 kgf m (7 – 10 lbf ft)
Nuts	2.2 – 2.8 kgf m (16 – 20 lbf ft)

40.4a "IN" mark should face to rear of engine

40.4b Fit piston, dowels and cylinder base gasket

40.5 Fit new O-ring to cylinder barrel spigot

40.7 Lower barrel into position as shown

40.8a Fit cam chain guide into recess

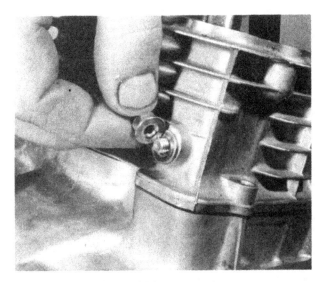

40.8b Secure tensioner with lower mounting nut

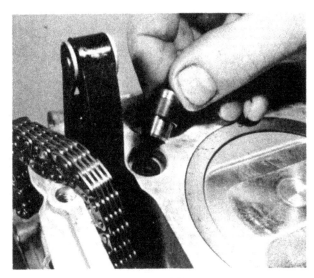

40.9a Note oil seal fitted around dowel pin

40.9b Postion dowels as indicated, fit head gasket

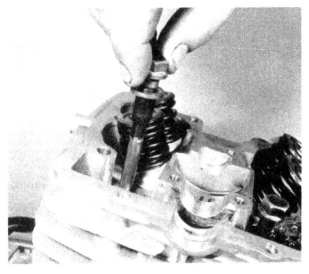

40.9c Drop cylinder head bolts into position...

40.9d ...and torque to specified setting

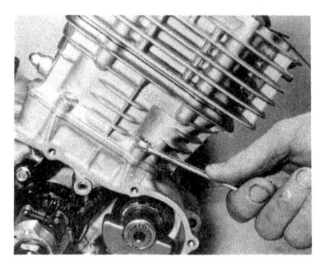

40.9e Do not omit to tighten cylinder base bolts

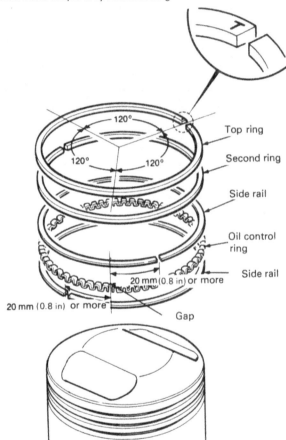

Fig. 1.20 Piston ring gap positions

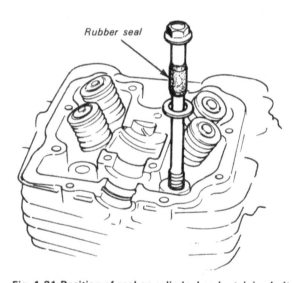

Fig. 1.21 Position of seal on cylinder head retaining bolt

41 Engine reassembly: fitting the camshaft and cylinder head cover – setting the valve timing

1   Check that the cylinder head area is clean and dry, then lubricate the cam bearing surface with molybdenum disulphide grease. Slide the camshaft into position from the right-hand side of the engine unit, looping the cam chain inside the sprocket flange. Offer up the camshaft sprocket with the two timing marks facing inwards, or towards the sparking plug.

2   Using a socket or box spanner passed through the access hole in the left-hand outer casing, turn the crankshaft anti-clockwise until the T mark aligns with the index mark in the upper inspection hole. Arrange the camshaft sprocket so that the timing marks run parallel to the cylinder head gasket face, then place the cam chain around the sprocket without rotating either the sprocket or the crankshaft.

3   Arrange the camshaft so that the lobes point in roughly the 4 o'clock and 8 o'clock positions. The sprocket mounting holes should now coincide and the first of the two retaining bolts can be fitted. Turn the crankshaft through one complete revolution until the second bolt can be fitted. Tighten each one to 1.7 – 2.3 kgf m (12 – 16 lbf ft). Check the timing accuracy by turning the crankshaft, again anti-clockwise, until the T mark aligns. Check that the cam sprocket timing marks align.

4   If the rocker arms and spindles were removed from the cyinder head cover, they should be refitted, lubricating the spindles and rocker arm bores with engine oil during assembly. Ensure that each rocker shaft is fitted in its original position and that the wave washer is positioned at the sparking plug end of each rocker. When fitting the retaining pins, position each shaft so that the cutout coincides with the stud hole through the

cylinder head cover. The shaft can be turned as required using the slot provided in the shaft end. New O-rings should be fitted to obviate any risk of oil leakage.

5   The jointing faces of the cylinder head cover and the cylinder head should be clean and dry, and the cylinder head face coated with a thin, even film of silicone rubber jointing compound. Care must be exercised when applying the compound to avoid any excess reaching the camshaft where it might obstruct oilways. To this end, the compound should not be applied to the jointing face immediately around the two bearing journals particularly on the two pillars of the centre bearing.

6   Fit the two locating dowels to the cylinder head, then lubricate the valves and camshaft with engine oil, filling the pocket in which the cam lobes run. Fit the camshaft end plug, coating its edge with jointing compound to ensure an oil-tight joint. Slacken the value adjuster locknuts and screws, then fit the cylinder head cover and securing bolts. Tighten the bolts gradually and evenly in a diagonal sequence to prevent warpage of the light alloy cover. The recommended torque setting is 1.0 – 1.4 kgf m (7 – 10 lbf ft).

7   Before refitting the inspection covers, adjust the valve clearances as follows. Turn the crankshaft anti-clockwise until the T mark on the alternator rotor registers with the fixed index mark in the inspection hole. The engine should be on the compression stroke (both valves closed), if not, turn the engine

through 360° and realign the T mark. Using feeler gauges of the appropriate size, set each adjuster to give the required clearance between it and the valve stem. When set correctly the feeler gauge should be a light sliding fit between the two components. The recommended valve clearances are as follows.

	Inlet	Exhaust
XR250 model	0.08 mm	0.10 mm
	(0.003 in)	(0.004 in)
All others	0.05 mm	0.10 mm
	(0.002 in)	(0.004 in)

When the valve clearances have been set, the inspection cover can be refitted, using new O-rings where necessary.

8   At this stage take up initial cam chain slack. Align the alternator T mark and position the engine on its compression stroke as described in paragraph 7. On 250 models slacken the tensioner locknut projecting from the centre of the cylinder barrel rear face. On 500 models slacken the hexagon-headed bolt positioned on the rear face of the cylinder head and the domed nut located below it, by 1½ – 2 turns. Allow the tensioner to tension the chain automatically and then tighten the fastener(s). Note that this procedure only serves to take up initial cam chain slack and the cam chain should be tensioned after the engine has been installed in the frame and is running, see Section 43.

41.1a Check oilway, fill pocket with oil

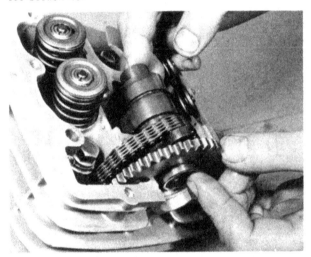

41.1b Assemble camshaft components as shown

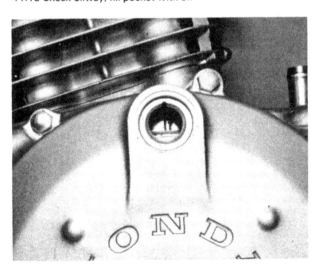

41.2 Valve timing is carried out with crankshaft 'T' mark aligned as shown

41.3 Secure camshaft sprocket

41.6a Place camshaft end plug into casing recess

Fig. 1.22 Valve timing camshaft alignment marks

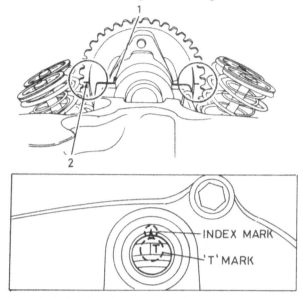

1   Cylinder head top face   2   Camshaft alignment mark

41.6b Fit cylinder head cover, not omitting hidden bolts

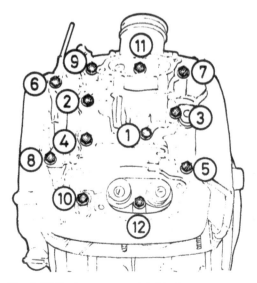

Fig. 1.23 Rocker cover nut tightening sequence

42 Engine reassembly: installing the rebuilt unit in the frame – final adjustments

1   As was the case during engine removal, installation can be completed by one person without insurmountable difficulties being encountered. It is helpful, however, to have some assistance whilst the unit is being lifted into position, as there will be less likelihood of the frame or paintwork sustaining damage. Lift the unit into the frame and assemble the front engine plates, the various mounting bolts, noting their varying lengths, and the cylinder head steady. Do not tighten any of the bolts until all are positioned loosely. The position of each bolt will be self evident and is indicated by its length. A mixture of 8 mm and 10 mm bolts are used on the 250 machines, the frame to engine plate connections being made by the smaller 8 mm bolts whilst 10 mm bolts are employed in all other instances. In the case of the 500 models the arrangement is identical except for the two rear mounting bolts which are of the 12 mm type. On all models, do not omit the spacer which is fitted on the left-hand side of the upper rear mounting bolt. When all of the bolts are in position they should be tightened to the appropriate torque setting shown below.

Engine mounting bolt torque settings
   XL250S, XR250
      8 mm bolts   2.0 – 3.5 kgf m (14 – 25 lbf ft)
      10 mm bolts   3.0 – 5.0 kgf m (22 – 36 lbf ft)
   XL500S, XR500
      8 mm bolts   2.0 – 3.5 kgf m (14 – 25 lbf ft)
      10 mm bolts   4.5 – 6.0 kgf m (33 – 43 lbf ft)
      12 mm bolts   7.0 – 10.0 kgf m (51 – 72 lbf ft)

2   Refit the left-hand footrest assembly ensuring that the locating ped engages in its slot in the frame, and secure it with its single retaining bolt. Assemble the right-hand footrest and brake pedal unit in a similar fashion, and reconnect the rear brake cable and brake switch operating spring. Check that the oil drain plug is tight, then fit the skid plate to the underside of the unit.

3   Manoeuvre the carburettor into position, fitting the throttle cables if these were released during removal. Set the lower cable adjusters to give approximately 2 – 6 mm of free play measured at the edge of the throttle twistgrip. Reconnect the

choke cable, ensuring that it is set up so that the choke is fully off when the knob is pushed home. Set the carburettor vertically between the intake and air cleaner hoses, then tighten the retaining clips.

4 Refit the final drive chain around the gearbox sprocket, positioning the latter on its shaft end. Fit a new lockwasher and the two retaining bolts. Tighten the bolts securely and lock them by bending up the washer tabs. Adjust the final drive chain tension, ensuring that the rear wheel is kept square in the swinging arm, then secure the rear wheel spindle. See Routine Maintenance for details on chain adjustment. Refit the sprocket cover and fit the gearchange pedal.

5 Connect the clutch cable to its operating arm, adjusting the cable to give 15 – 25 mm ($\frac{5}{8}$ – 1 in) free play measured at the lever end. Refit the decompressor cable between the cylinder head cover and the right-hand outer cover. The cable should be adjusted as described below noting that this operation must be carried out **after** the valve clearances have been set.

6 Using the inspection hole provided in the centre of the left-hand outer cover, turn the crankshaft anti-clockwise until the engine is on the compression stroke. This will be indicated by a high resistance to movement being felt. Carry on turning the crankshaft until the TDC (top dead centre) is reached, this being indicated in the upper inspection hole by the T mark coinciding with the index mark. Check the amount of free play at the tip of the decompressor lever on the cylinder head cover. This should be set using the cable adjuster to give 1 – 3 mm (0.04 – 0.12

in) free play in the case of the 250 machines or 1 – 2 mm (0.04 – 0.08 in) for the 500 models.

7 Trace the ignition and alternator leads, routing them along the frame tubes to their corresponding connector halves. Once connected the cables should be secured to the frame tubes using new cable ties, or in an emergency, pvc tape. Assemble the exhaust pipes using new exhaust port sealing rings to prevent leakage. Fit the split rollers and retainers at the exhaust ports and tighten the clamp at the silencer joint.

8 On XL models, fit and reconnect the battery and check that the electrical system functions properly. Note that the headlamp operates directly from the alternator and thus can only be checked when the engine is running. This applies to the entire electrical system on the XR models.

9 Refit the side panels, fuel tank and seat noting that the seat strap, where fitted, should be secured by the bolts in the rear suspension unit mountings. Connect the fuel feed pipe between the tank and carburettor. Remove the oil filler plug and add 1.5 litres (1.6 US, quarts/2.6 Imp pints) of SAE 10W – 40 motor oil. Hold the machine upright and rest the combined filler plug and dipstick on the top of the filler hole. Withdraw the dipstick and check that the oil level is within the upper and lower limits. Note that the engine oil level, must be rechecked soon after the engine has been run, because a small amount of topping up may be necessary after the oil has been distributed around the engine and transmission components.

42.1a The engine front mounting plate

42.1b The cylinder head mounting plates in position

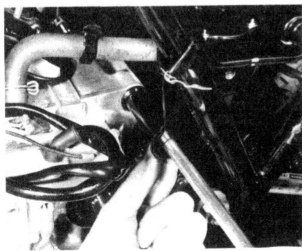

42.1c Slide upper rear bolt through spacer and casing

42.1d Fit lower rear bolt as shown

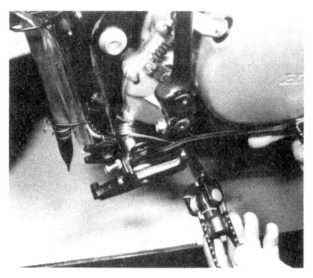

42.2a Right-hand footrest assembly is fitted as shown...

42.2b ...and retained by a single bolt

42.2c Skid plate is retained to front engine plate...

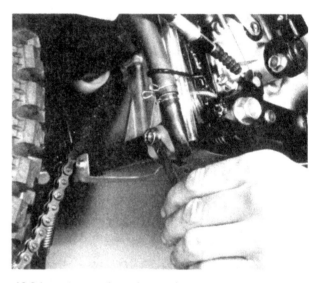

42.2d ...and to rear frame brace tube

42.3 Reconnect cables and install carburettor

42.4a Fit chain and sprocket to shaft and...

42.4b ...and secure with retainer plate

42.7a Use new exhaust port sealing rings

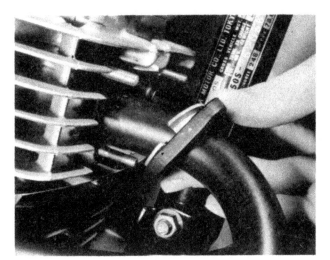

42.7b Pipe is retained by collet halves and flange

43 Starting and running the rebuilt engine

1 Turn on the fuel tap, close the choke, and attempt to start the engine by means of the kickstart pedal. Do not be disillusioned if there is no sign of life initially. A certain amount of perseverance may prove necessary to coax the engine into activity even if new parts have not been fitted. Should the engine persist in not starting, check that the spark plug has not become fouled by the oil used during re-assembly.Failing this go through the fault finding charts and work out what the problem is methodically.

2 When the engine does start, keep it running as slowly as possible to allow the oil to circulate. Open the choke as soon as the engine will run without it. During the initial running, a certain amount of smoke may be in evidence due to the oil used in the reassembly sequence being burnt away. The resulting smoke should gradually subside.

3 Check the engine for blowing gaskets and oil leaks. Before using the machine on the road, check that all the gears select properly, and that the controls function correctly.

4 Soon after starting the engine for the first time the cam chain tension must be adjusted. Allow the engine to run at tick-over (1200 ± 100 rpm). On 250 models loosen the locknut located to the rear of the cylinder barrel, immediately below the carburettor, and then tighten the nut. Whilst the nut is loose the tensioner will automatically tension the chain correctly. Adjustment on 500 models is similar except that there are two tensioner fasteners, a bolt and a domed nut. Loosen both between 1½ and 2 turns and then tighten them again. Do not loosen either fastener by more than 2 turns or the tensioner may become displaced, allowing engine damage to occur.

44 Fault diagnosis: engine

Symptom	Cause	Remedy
Engine will not start	Defective sparking plug	Remove the plug and lay it on cylinder head. Check whether spark occurs when ignition is switched on and engine rotated.
	Dirty or closed contact breaker points	Check condition of points and whether gap is correct.
	Faulty or disconnected condenser	Check whether points arc when separated. Replace condenser if evidence of arcing.

Engine runs unevenly	Ignition and/or fuel system fault	Check each system independently, as though engine will not start.
	Blown cylinder head gasket	Leak should be evident from oil leakage where gas escapes.
	Incorrect ignition timing	Check accuracy and if necessary reset.
Lack of power	Fault in fuel system or incorrect ignition timing	See above
Heavy oil consumption	Cylinder barrel in need of rebore	Check for bore wear, rebore and fit oversize piston if required.
	Damaged oil seals	Check engine for oil leaks.
Excessive mechanical noise	Worn cylinder barrel (piston slap) Worn big end bearings (knock) Worn main bearings (rumble)	Rebore and fit oversize piston Fit replacement crankshaft assembly. Fit new journal bearings and seals.
Engine overheats and fades	Lubrication failure	Stop engine and check whether internal parts are receiving oil. Check oil level in crankcase.

45 Fault diagnosis: clutch

Symptom	Cause	Remedy
Engine speed increases as shown by tachometer but machine does not respond	Clutch slip	Check clutch adjustment for free play at handlebar lever. Check thickness of inserted plates.
Difficulty in engaging gears. Gear changes jerky and machine creeps forward when clutch is withdrawn. Difficulty in selecting neutral	Clutch drag	Check clutch adjustment for too much free play. Check clutch drums for indentations in slots and clutch plates for burrs on tongues. Dress with fine file if damage not too great.
Clutch operation stiff	Damaged, trapped or frayed control cable	Check cable and replace if necessary. Make sure cable is lubricated and has no sharp bends.

46 Fault diagosis – gearbox

Symptom	Cause	Remedy
Difficulty in engaging gears	Selector forks bent Gear clusters not assembled correctly	Replace. Check gear cluster arrangement and position of thrust washers.
Machine jumps out of gear	Worn dogs on ends of gear pinions Stopper arms not seating correctly	Replace worn pinions. Remove right hand crankcase cover and check stopper arm action.
Gearchange lever does not return to original position	Broken return spring	Replace spring.
Kickstarter does not return when engine is turned over or started	Broken or poorly tensioned return spring	Replace spring or re-tension.
Kickstarter slips	Ratchet assembly worn	Part crankcase and replace all worn parts.

Chapter 2 Fuel system and lubrication

For modifications and information relating to later models, see Chapter 7

Contents

Specifications

	XL250 S	XR250, XL500 and XR500
Fuel tank		
Overall capacity	9.5 lit (2.5/2.1 US/Imp gal)	10 lit (2.6/2.2 US/Imp gal)
Reserve capacity	2.0 lit (0.5/0.4 US/Imp gal)	2.0 lit (0.5/0.4 US/Imp gal)

Carburettor	XL250 S	XR250	XL500 S	XR500
Make	Keihin	Keihin	Keihin	Keihin
Type	PD 04B	PD 02A	PD 08A	PD 06A
Venturi size	28 mm (1.10 in)	–	32 mm (1.26 in)	–
Main jet	120	122	155	158
Pilot (slow) jet	40	40	58	55
Pilot screw setting (No turns out)	1⅝	2¼	2	2
Jet needle clip position	–	4th groove	–	3rd groove
Idle speed	1200 ± 100 rpm	1300 ± 100 rpm	1200 ± 100 rpm	1200 ± 100 rpm
Float level	14.5 mm (0.57 in)	14.5 mm (0.57 in)	14.5 mm (0.57 in)	14.5 mm (0.57 in)

Lubrication system

All models

Type ....
Wet sump, forced lubrication to major engine/gearbox components

Filtration ....
Gauze strainer

Oil pump

Type ....
Trochoidal

Rotor to body clearance ....
0.15–0.18 mm (0.006–0.007 in)

Service limit ....
0.25 mm (0.010 in)

Inner to outer rotor clearance ....
0.15 mm (0.006 in)

Service limit ....
0.20 mm (0.008 in)

Axial clearance ....
0.01–0.07 mm (0.0004–0.0028 in)

Service limit ....
0.1 mm (0.004 in)

1   General description

The models covered by this manual are each equipped with a simple lever-operated slide carburettor. Fuel is fed by gravity from the fuel tank via a three position fuel tap. The three tap positions are 'Off', 'On' and 'Reserve', the latter providing an emergency supply of fuel and a warning that the fuel level has run low. Fuel from the tank enters the float chamber of the carburettor where the level is kept constant by a float-operated valve. The fuel/air mixture is controlled by the throttle valve, the needle/needle jet arrangement and by the main jet and pilot (slow) jet. A diaphragm air cut-off valve is fitted to all models. This device reduces the amount of air entering the exhaust system during over-run, thus reducing the problem of backfiring induced by the weak mixture being ignited by the hot exhaust pipe.

The US, Canadian and certain European market models are fitted with an accelerator pump (XL 250S model only) to compensate for the otherwise weak mixture required by these countries. An adjustable fast-idle device is also fitted.

A comprehensive force-fed lubrication system is employed. A trochoid oil pump draws oil from the sump via a small gauze strainer which traps any metal particles. At the pump outlet the oil splits into a number of circuits. One of these feeds oil through a spring-loaded quill into the right-hand end of the crankshaft. The oil is routed through the crankshaft and flywheel to the big-end bearing, the escaping oil providing splash lubrication for the small-end bearing and cylinder wall before returning to the sump.

A branch off the crankshaft feed supplies oil to the camshaft and rockers and also to the front balancer shaft. Oil is also fed under pressure to the gearbox layshaft and mainshaft assemblies.

2   Fuel tank: removal, examination and replacement

1   To gain access to the single tank-retaining bolt it is first necessary to release the seat. This is secured by a bolt on each side of the seat base, plus the seat strap on XL versions. The strap is retained by a bolt which screws into the suspension unit mounting on each side of the machine.

2   The seat can now be lifted clear to reveal the single retaining bolt which passes through a lug into the frame. Slacken and remove the bolt, then lift the rear of the tank upwards. This will make access to the fuel pipe easier, allowing it to be prised off once the tap has been turned to the 'Off' position. Once removed, the tank can be pulled back to free the front mounting rubbers and lifted away.

3   If the tank is to be stored it should be placed in a safe place away from any area where fire is a hazard and where the paint finish may become damaged.

4   Any signs of fuel leakage should be dealt with promptly in view of the risk of fire or explosion should fuel drip onto the hot exhaust system. It is **not** recommended that the tank is repaired using welding or brazing techniques, because even a small amount of residual fuel vapour can result in a dangerous explosion. A more satisfactory alternative is to use one of the resin-based tank sealing compounds. These are designed to line the tank with a tough fuel-proof skin, sealing small holes or splits in the process. The suppliers of these products advertise regularly in the motorcycle press.

5   Tank fitting is a straightforward reversal of the removal sequence. If problems are encountered when attempting to engage the front mounting rubbers, a trace of petrol will act as a lubricant, easing assembly appreciably. After the fuel pipe has been refitted, turn the fuel on and check for leaks.

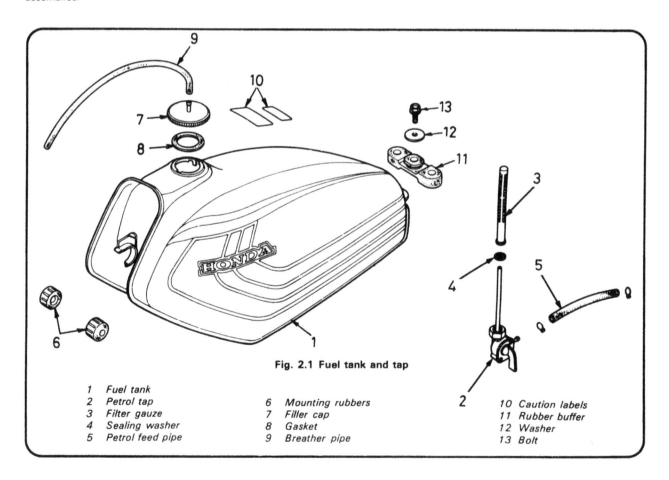

Fig. 2.1 Fuel tank and tap

1	Fuel tank	
2	Petrol tap	6 Mounting rubbers 10 Caution labels
3	Filter gauze	7 Filler cap 11 Rubber buffer
4	Sealing washer	8 Gasket 12 Washer
5	Petrol feed pipe	9 Breather pipe 13 Bolt

3 Petrol feed pipe: examination

1 The petrol feed pipe is made from thin walled synthetic rubber and is of the push-on type. It is necessary to replace the pipe only if it becomes hard or splits. It is unlikely that the retaining clips will need replacing due to fatigue as the main seal between the pipe and union is effected by an 'interference' fit.

2 If the petrol pipe has been replaced with the transparent plastic type for any reason, look for signs of yellowing which indicates that the pipe is becoming brittle due to the plasticiser being leached out by the petrol. It is a sound precaution to renew a pipe when this occurs, as any subsequent breakage whilst in use will be almost impossible to repair.

Note: On no account should natural rubber tubing be used to carry petrol, even as a temporary measure. The petrol will dissolve the inner wall, causing blockages in the carburettor jets which will prove very difficult to remove.

4 Petrol tap: removal, examination and replacement

1 Before the petrol tap can be removed, it is first necessary to drain the tank. This is easily accomplished by removing the feed pipe from the carburettor float chamber and allowing the contents of the tank to drain into a clean receptacle, with the tap turned to the 'reserve' position. Alternatively, the tank can be removed and place d on one side, so that the fuel level is below the tap outlet. Take care not to damage the paintwork.

2 The tap unit is retained by a gland nut to the threaded stub on the underside of the tank. It can be removed after the fuel pipe has been pulled off the tap.

3 If the tap lever leaks, it will be necessary to renew it as a complete unit. It is not possible to dismantle the tap for repair.

4 When reassembling the tap, reverse the procedure for dismantling.

5 Check that the feed pipe from the tap to the carburettor is in good condition and that the push-on joints are a good fit, irrespective of the retaining wire clips. If particles of rubber are found in the filter, replace the pipe, since this is an indication that the internal bore is breaking up.

6 If there have been indications of water contamination in the fuel, the removal of the tap presents a good opportunity to drain and flush the tank completely. Many irritating fuel system faults can be traced to water in the petrol. This often appears as a result of condensation inside the petrol tank. The resulting blobs of water are easily drawn into the carburettor, where they can cause intermittent blockages in the jets and drillings. Any accumulations of water should therefore be flushed from the tank before the tap is refitted. The tubular filter gauze should be removed and cleaned carefully prior to reassembly.

5 Carburettor: removal

1 To gain access to the carburettor for removal purposes it is first necessary to remove the seat and fuel tank, as described in Section 2, and to pull off the side panels. Slacken the locknuts which retain the throttle cable adjusters to the anchor plate. The adjusters can now be displaced and the inner cables disengaged from the operating pulley. Slacken the choke cable clamping screw and release the choke cable from the carburettor.

2 Slacken the retaining clips which secure the carburettor to the inlet adaptor and the air cleaner hose. The carburettor can now be pulled back to free it from the adaptor and manoeuvred clear of the engine.

6 Carburettor: dismantling and reassembly

1 Start by draining the residual fuel from the float bowl by means of the small drain screw which screws into the base of the float bowl. Slacken and remove the two screws which retain the carburettor top and lift the top away. Release the nut and washer on the end of the throttle spindle and remove the fast idle link, taking care to acoid straining the accelerator pump spring (US and Canadian XL models only). With the remaining models, remove the small screw which retains the throttle link to the spindle.

2 Unhook the throttle return spring, and withdraw the quadrant and spindle. The throttle lever will now be freed and should be withdrawn together with the throttle valve assembly. The lever and throttle valve can be separated after releasing the connecting link which joins them. This is accomplished by removing the small tension spring which retains the pivot of the lever and the throttle valve bracket to the link. To complete the dismantling of the throttle valve assembly, remove the two small screws which secure the bracket to the throttle valve, then lift away the bracket, spring and jet needle.

3 The air cut-off valve is located behind a circular cover on the side of the instrument and is retained by two screws. Remove the screws and lift away the cover and spring to expose the diaphragm assembly. This is a fragile component, and care must be taken to avoid damage as it is removed. Peel away the edges of the diaphragm, then remove it together with the brass valve plunger.

4 The accelerator pump, where fitted, consists of a rod-operated diaphragm unit similar in construction to the air cut-off valve. The operating rod passes vertically upwards to a point just below the throttle lever pivot. A pump lever is connected by a spring to the fast idle link. The latter turns with the throttle pulley thus applying pressure to the pump lever. This in turn depresses the pump rod causing a metered quantity of fuel to be injected as the throttle is opened. To dismantle the pump components, release the three screws which retain the pump cover, lifting this and the return spring away. Carefully remove the pump diaphragm and withdraw it together with the pump rod. The pump lever pivots on a hexagon and cross-headed bolt and can be removed when this has been unscrewed.

5 Remove the three float bowl retaining screws, then lift the bowl away, taking care not to damage the sealing ring. The float can be removed after displacing and withdrawing its pivot pin. The float needle will come away with the float to which it is attached by a fine wire loop.

6 The main jet is located at the centre of the carburettor and is identified by its slotted cheese head. It can be unscrewed on its own or together with the hexagon headed needle jet holder to which it is attached. It should be noted that the adjacent projection is the pilot (slow) jet. This is pressed into the carburettor body and cannot be removed. Carefully screw the pilot mixture screw inwards, counting the number of turns or part turns so that it can be fitted in its original setting. Do not screw it hard against its seating.

7 Reassembly is tackled in the reverse of the dismantling order, using new seals and O-rings as required. Each part must be scrupulously clean, and care must be exercised to avoid overtightening any of the carburettor components. All of these are rather delicate and can easily become damaged. After reassembly, the various carburettor adjustments should be checked as described later in this Chapter.

8 On 1980 US models a pilot screw limiter cap is fitted to the pilot screw to prevent injudicious adjustment which might cause mixture changes which exceed the EPA regulations. If the screw is removed it must be refitted prior to installation of the float bowl, and placed in precisely the same position as it was in prior to removal. If a new screw is to be fitted adjustment should be carried out as described in Section 8, and a new limiter cap glued into place after completion of adjustment.

U.S.TYPE ONLY

U.S.TYPE ONLY

Fig. 2.2 Carburettor

1 Carburettor body	35 Bolt
2 Carburettor top	36 Washer – 2 off
3 Gasket	37 Spring washer
4 Screw – 2 off	38 Return spring
5 Throttle link	39 Float bowl
6 Jet needle	40 Sealing ring
7 Screw	41 Float
8 Spring washer	42 Pivot pin
9 Return spring	43 Float needle
10 Washer	44 Wire loop
11 Throttle valve bracket	45 Screw – 3 off
12 Screw	46 Float bowl drain screw
13 Spring	47 Anti-surge baffle
14 E-clip	48 Main jet
15 E-clip	49 Needle jet holder
16 Washer	50 Needle jet
17 Throttle valve	51 Pilot mixture screw
18 Throttle pump arm	52 Spring
19 Choke cable anchor	53 Washer
20 Fast idle adjuster	54 O-ring
21 Spring	55 Air cut-off valve cover
22 Nut	56 Diaphragm
23 Spring washer	57 Spring
24 Nut	58 Hose guide
25 Rubber boot	59 Screw – 2 off
26 Cable clamp	60 O-ring
27 Screw – 2 off	61 Drain pipe
28 Throttle cable bracket	62 Clip – 2 off
29 Screw	63 Accelerator pump assembly
30 Throttle stop screw	64 Rod/diaphragm
31 Spring	65 Spring
32 Throttle cable spring	66 Pump cover
33 Return spring	67 Screw – 3 off
34 Accelerator pump arm	

6.1a Carburettor top is secured by two screws

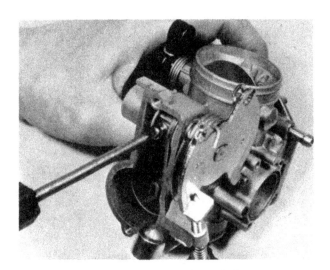

6.1b Slacken screw which retains link to spindle

4.2 Fuel tap is retained to tank by a gland nut

6.2a Throttle assembly can be withdrawn from carburettor

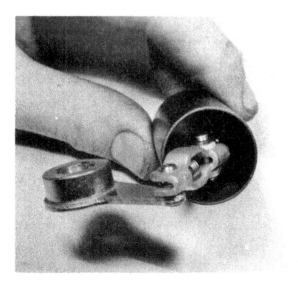

6.2b Link assembly is screwed to throttle valve

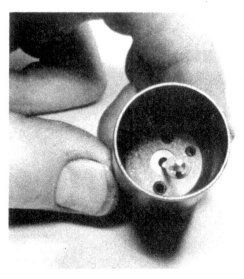

6.2c Needle is located by clip

6.3a Remove air cut-off valve cover and spring

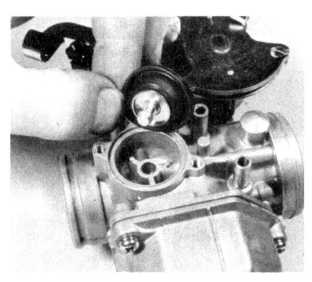

6.3b Valve diaphragm and plunger can be lifted away

6.3c Note small O-ring in recess (arrowed)

6.5a Float bowl is secured by three screws

6.5b Displace the float pivot pin...

6.5c ...and remove float together with needle

6.6a Note position of cut-out in main jet baffle

6.6b Main jet can be unscrewed for cleaning...

6.6c ...as can hexagon-headed needle jet

6.6d Note pilot screw setting prior to removal

7 Carburettor: examination and renovation

1   Having dismantled the carburettor as described in Section 6, the various components should be laid out for examination. If symptoms of flooding have been in evidence, check that the float is not leaking, by shaking and listening for petrol inside. It is rare to find leaks in plastic floats, this problem being more common in the brass type.

2   A more likely cause of flooding is dirt on the float needle or its seat. The needle can easily be removed, with the float detached, by inverting the body and allowing it to drop, from its seating. Examine the faces of the needle and seat for foreign matter and also for scoring. If in bad condition, renew the needle and note whether any improvement is obtained. The valve seat cannot be removed from the body and if badly damaged the entire body must be renewed. The only alternative is to get the seat area refaced by a light engineering company, remembering that this will upset the float height setting.

3   The main jet screws into the needle jet, which is central in the carburettor body. It is not prone to any real degree of wear, but can become blocked by contaminants in the petrol. These can be cleared by an air jet, either from an air lne or a foot pump. As a last resort, a fine bristle fron a nailbrush or similar may be used, but on no account should wire be used as this may damage the precision drilling of the jet.

4   The needle jet may become worn after a considerable mileage has been covered and should be renewed along with the needle. Always fit replacement parts as a pair.

5   The pilot jet is located adjacent to the main jet and needle jet assembly. It is pressed into position and thus must be cleaned in situ.

6   Examine the throttle valve for scoring or wear, renewing if badly damaged. If damage is evident, check the internal bore of the carburettor, and if necessary renew this also. Check that the needle is free from scoring or other damage, and roll it on a flat surface to check that it has not become bent.

8 Carburettor: adjustments and settings

1   The various jet sizes, throttle valve cutaway and needle poisiton are predetermined by the manufacturer and should not require modification. Check with the Specifications list at the beginning of this Chapter if there is any doubt about the types fitted.

2   Before any attempt at adjustment is made, it is important to understand which parts of the instrument control which part of its operating range. A carburettor must be capable of delivering

the correct fuel/air ratio for any given engine speed and load. To this end, the throttle valve, or slide as it is often known, controls the volume of air passing through the choke or bore of the instrument. The fuel, on the other hand, is regulated by the pilot and main jets, by the jet needle, and to some extent, by the amount of cutaway on the throttle valve.

3   As a rough guide, up to $\frac{1}{8}$ throttle is controlled by the pilot jet, $\frac{1}{8}$ to $\frac{1}{4}$ by the throttle valve cutaway, $\frac{1}{4}$ to $\frac{3}{4}$ throttle by the needle position and from $\frac{3}{4}$ to full by the size of the main jet. These are only approximate divisions, which are by no means clear cut. There is a certain amount of overlap between the various stages.

4   If any particular carburation fault has been noted, it is a good idea to try to establish the most likely cause before dismantling or adjusting takes place. If, for example, the engine runs normally at road speeds, but refuses to tick over evenly, the fault probably lies with the pilot mixture system, and will most likely prove to be an obstructed jet. Whatever the problem may appear to be, it is worth checking that the jets are clear and that all the components are of the correct type. Having checked these points, refit the carburettor and check the settings as follows.

5   Set the pilot mixture screw to the position given in the Specifications Section. Start the engine, and allow it to attain its normal working temperature. This is best done by riding the machine for a few miles. Set the throttle stop screw to give a normal idling speed. Try turning the pilot mixture screw inwards by about $\frac{1}{4}$ turn at a time, noting its effect on the idling speed, then repeat the process, this time bringing the screw outwards. The pilot mixture screw should be set in the position which gives the the fastest consistent tickover. If desired, the tickover speed may be reduced further by lowering the throttle stop screw, but care should be taken that this does not cause the engine to falter and stop after the throttle twistgrip has been opened and closed a few times.

6   It is important to note that in the case of US models and models supplied to certain other markets, the carburettor is set up to meet the stringent emission laws which prevail in these areas. No unauthorised adjustments are permitted in these instances, and great care should be taken to avoid carrying out any alteration which might cause an abnormally high toxic gas emission rate. If in doubt, check with local regulations before attempting any adjustment.

7   On XR models, a reduction in main jet size is recommended if the machine is used at high altitudes. The standard main jet can be used safely up to an altitude of 6500 ft (2000 metres) above which a No 118 (XR250) or No 152 (XR 500) main jet should be fitted. The smaller jets will compensate for the lower air density experienced as altitude is gained. These jets should

7.2 Examine float needle tip for wear or damage

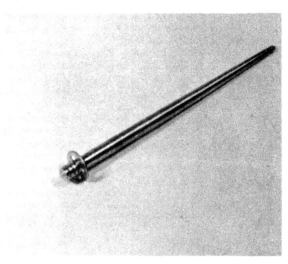

7.6 Check needle for straightness and scoring

not be used below 5000 ft (1500 metre) because the weak mixture that results will cause overheating and possible engine damage. When changing the main jet an alteration in pilot mixture strength will probably be necessary. The above remarks apply equally to XL models, although no specific jet change recommendations are given by the manufacturer. It is suggested that similar jet reductions are made proportional to the normal main jet size.

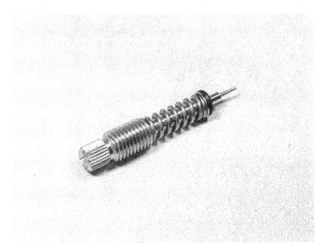

8.5 Pilot screw tip is easily damaged

9 Carburettor: checking and setting the float height

1   It is important that the level of fuel in the float bowl in maintained at the prescribed height to avoid adverse effects on the mixture strength. It is worth noting that unless the correct float height is set, it will be impossible to set the remaining adjustments to obtain efficient running.

2   The float height is measured between the gasket face of the carburettor body and the bottom of the float. This should be done with the carburettor turned 90° to its normal position so that the weight of the float is not applied to the valve needle. The latter should just bear upon the valve seat when the measurement is made. In all cases, the correct height should be 14.5 mm (0.57 in).

3   If adjustment is required it should be made by carefully bending the small tang which operates the valve needle. Note that a very small amount of movement at the tang will translate into a much larger change in the float height, so make any adjustments as fine as possible.

10 Air filter: general description and maintenance

1   The air used in combustion is drawn into the carburettor via an air filter element. This performs the vital job of removing dust and any other airborne impurities which would otherwise enter the engine, causing premature wear. It follows that the element must be kept clean and renewed, if damaged, as it will have an adverse effect on performance if neglected. Apart from the obvious problem of increased wear caused by a damaged element, a clogged or broken filter will upset the mixture setting, allowing it to become too rich or too weak.

2   The filter element is housed in a moulded plastic casing beneath the seat and is accessible via the left-hand side panel. Thje latter is retained by three screws. Once the panel has been released, the element assembly can be removed by pulling the retaining spring outwards (250 models) or by releasing the wing nut and retaining clip (500 models).

3   The air cleaner element on all models consists of a band of foam rubber supported by an expanded metal frame. The element is impregnated with oil to trap any dust passing through it. The foam may be cleaned by detaching it from its frame and washing it in a high flash-point solvent – do not use petrol (gasoline) because of the risk of fire. Dry the element thoroughly, then soak it in new SAE 80 or 90 gear oil, squeezing it to remove any excess oil. The element may now be refitted.

4   It is difficult to give any firm recommendations about cleaning intervals, as a trail bike or enduro machine is used for widely differing purposes. Again experience must be the best guide, and cleaning may be necessary weekly, or at up to 3000 mile intervals (500 km). The element will eventually become degraded to the point that renewal is necessary, and obviously must be renewed if it becomes torn or holed.

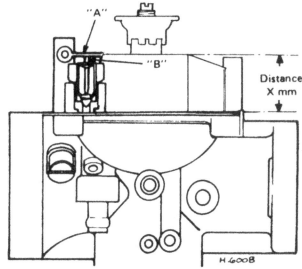

Fig. 2.3 Checking the float height

A   Float tongue
B   Float valve
X   Float height

10.2a Side cover is retained by three screws

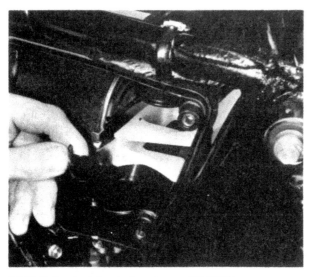

10.2b Pull out retainer to free element (XL250S)

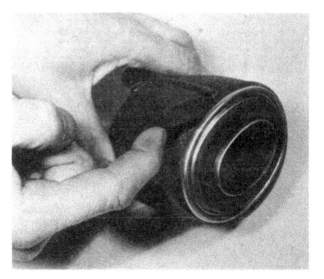

10.3 Peel element off frame for cleaning

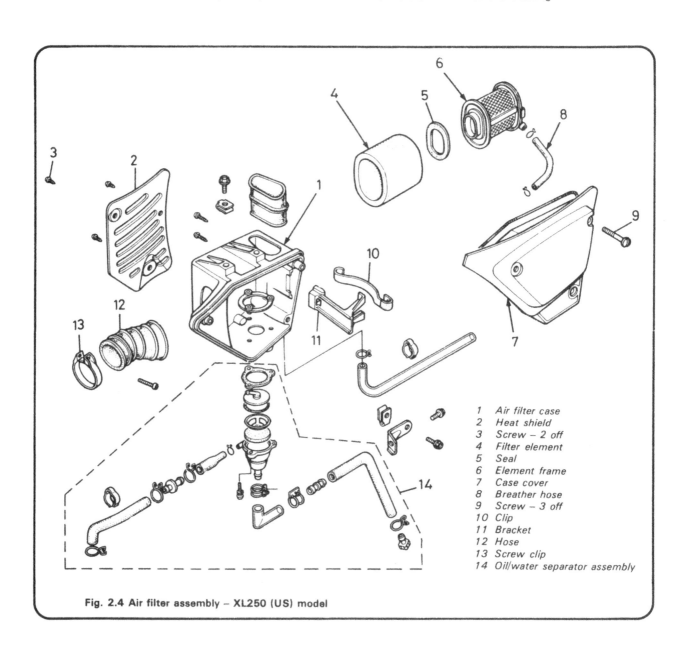

1   Air filter case
2   Heat shield
3   Screw – 2 off
4   Filter element
5   Seal
6   Element frame
7   Case cover
8   Breather hose
9   Screw – 3 off
10 Clip
11 Bracket
12 Hose
13 Screw clip
14 Oil/water separator assembly

Fig. 2.4 Air filter assembly – XL250 (US) model

Fig. 2.5 Air filter assembly – XL250S

1   Air filter case
2   Element frame
3   Element
4   Seal
5   Left-hand side panel
6   Seal
7   Hose
8   Screw – 3 off
9   Duct
10 Heat shield
11 Clip
12 Bracket
13 Hose
14 Oil/water separator –
    US models
15 Breather assembly –
    UK models
16 Hose
17 Hose
18 Hose
19 Clamp
20 Screw
21 Hose clip
22 Spring clip
23 Union
24 Drain valve
25 T-piece
26 Hose

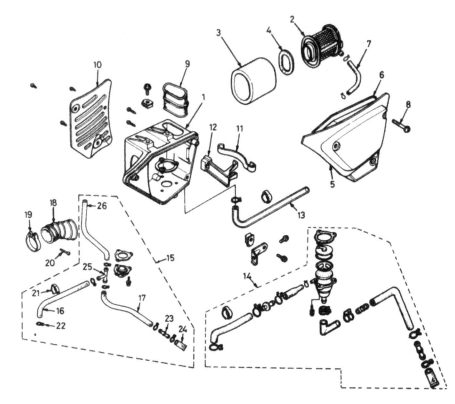

1   Air filter case
2   Duct
3   Hose
4   Filter element
5   Element frame
6   Frame mounting plate
7   Washer – 2 off
8   Screw – 2 off
9   Screw
10 Washer
11 Wing nut
12 Clamp
13 Screw
14 Washer
15 Drain valve
16 Spring clip
17 Seal
18 Bracket
19 Screw
20 Hose
21 Screw
22 Spring clip
23 Drain plug
24 Screw
25 Clamp
26 Hose
27 Spring clip –
    5 off
28 T-piece
29 Union
30 Drain valve
31 Hoseclip
32 Hose

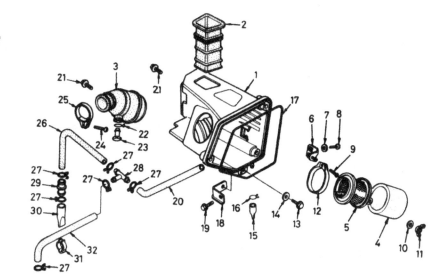

Fig. 2.6 Air filter assembly – XR500

11 Exhaust system: general description and maintenance

1   The exhaust system on all models comprises a pair of exhaust pipes, which terminate in a single, large diameter tube at the silencer joint, and a single silencer. The silencer is mounted on the right-hand side of the machine, running behind the right-hand side panel and below the seat. This arrangements keeps the system well out of harm's way, but does require a fair amount of dismantling work should the need for removal arise.

2   The exhaust pipes are reasonably easy to remove in all cases. Start by releasing the exhaust port retainers and split collets, each of which is retained by two nuts. Slacken the exhaust pipe clamp at the silencer junction. The pipes can now be pulled forward and lifted clear of the machine.

3   If it proves necessary to remove the silencer from the frame, start by detaching the seat from the frame by removing its two mounting bolts and, where fitted, the seat strap. Remove the right-hand side cover and, if not already removed, the exhaust pipes. From this point on, the procedure applicable to the 250 models differs from that employed on the 500 machines.

XL250S and XR250

4   Remove the front section of the rear mudguard by pulling it clear of the frame tubes to which it clips. Release the left-hand side cover and air cleaner lid. Remove the two screws which retain the top of the air cleaner casing to the frame, then release the single lower fixing screw and slacken the hose clip at the carburettor intake. The air cleaner assembly should now be pulled back and manoeuvred clear of the frame. Make a careful note of the position of the breather tube arrangement as a guide during reassembly.

5   Slacken and remove the rear silencer clamp, and release the central fixing bolt which passes down through the frame, noting the disposition of the heat insulating washer. This must be retained and fitted in the correct position during reassembly. Release the front fixing nut which also serves to retain the clutch cable anchor plate, again noting the position of the heat insulating washers fitted on either side of the plate. The silencer unit can now be threaded out of the frame. Pull the unit back, turning it so that the front end can be passed out through the space normally occupied by the air cleaner assembly.

XL500S and XR500

6   Silencer removal on the larger models is somewhat easier in view of the less bulky construction of the main silencer box. It will be necessary to detach the front section of the mudguard which is clipped to the frame tubes. The air cleaner casing can be left undisturbed. The silencer unit is secured by three retaining bolts in a similar manner to that described for the smaller models. Note that the centre mounting bolt passes horizontally through a bracket and the front mounting employs a bolt rather than a stud and nut.

7   Reassembly is a reversal of the appropriate removal sequence, noting that the entire system should be in place before the various bolts are tightened. This will avoid any alignment problems between the two halves of the system. Ensure that the insulating washers are fitted to avoid heat being transmitted to the frame, and check for exhaust leaks when assembly is complete.

8   The exhaust system on a four-stroke motorcycle will require very little attention, because, unlike two-stroke machines, it is not prone to the accumulation of carbon. The only points requiring attention are the general condition of the system, including mountings and the surface finish and ensuring that the system is kept airtight , particularly at the exhaust port. Air leaks here will cause mysterious backfiring when the machine is on overrun, as air will be drawn in causing residual gases to be ignited in the exhaust pipe. To this end, make sure that the composite sealing ring is renewed each time the system is removed.

9   On most road going machines, the exhaust system is protected to some extent by its chromium plated finish. The matt black finish employed on the Honda XL and XR models may prove rather less durable, in view of the off-road use to which the machine is designed to be put. It can be touched up, where and when necessary, with one of the specially designed aerosol spray paints such as Hermetite Heat Resistant Paint. These are capable of surviving exhaust system temperatures, and are often known as VHT coatings, indicating very high temperature.

10 The XR models have small detachable plates screwed to the silencer. These are fitted to allow cleaning. Remove the plates and start the engine. The exhaust gases will expel any accumulated carbon, and this process should be continued until all loose debris has been ejected. Needless to say, this procedure should be carried out in a well ventilated area and care should be taken not to stand in the path of debris ejected from the exhaust system.

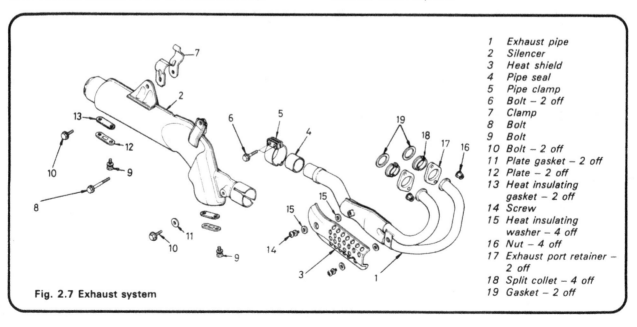

Fig. 2.7 Exhaust system

1   Exhaust pipe
2   Silencer
3   Heat shield
4   Pipe seal
5   Pipe clamp
6   Bolt – 2 off
7   Clamp
8   Bolt
9   Bolt
10 Bolt – 2 off
11 Plate gasket – 2 off
12 Plate – 2 off
13 Heat insulating gasket – 2 off
14 Screw
15 Heat insulating washer – 4 off
16 Nut – 4 off
17 Exhaust port retainer – 2 off
18 Split collet – 4 off
19 Gasket – 2 off

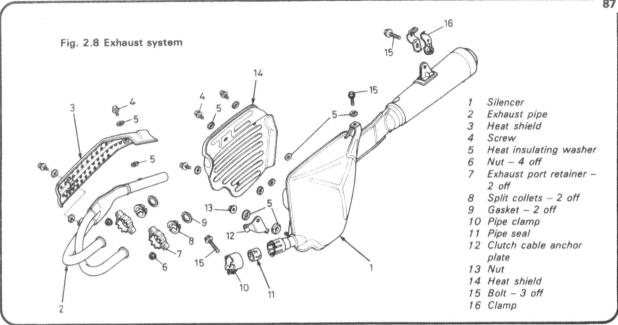

Fig. 2.8 Exhaust system

1 Silencer
2 Exhaust pipe
3 Heat shield
4 Screw
5 Heat insulating washer
6 Nut – 4 off
7 Exhaust port retainer –
  2 off
8 Split collets – 2 off
9 Gasket – 2 off
10 Pipe clamp
11 Pipe seal
12 Clutch cable anchor
  plate
13 Nut
14 Heat shield
15 Bolt – 3 off
16 Clamp

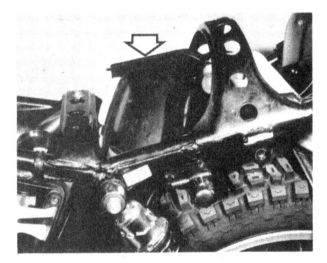

11.4a Remove front section of rear mudguard (arrowed)

11.4b Air filter casing is held by two screws at top...

11.4c ...and single lower screw (arrowed)

11.4d Casing is removed to give access to silencer

12 Oil pump: removal, examination and reassembly

1 The oil pump will not normally require specific attention until a complete engine overhaul is necessary. If, however, it proves necessary to examine the pump for any reason, it can be removed from the crankcase after detaching the right-hand outer cover and the clutch. Removal of the pump is covered specifically in Section 10 of Chapter 1. The various pump components should be checked for wear or damage as follows.

2 Examine each component for signs of scuffing and wear. Note especially the condition of the rotors and pump body. If these are at all worn, the pump must be renewed. Replace the spindle and rotors, and measure the clearance between the outer rotor and the pump body using feeler gauges. The nominal clearance is 0.15 – 0.18 mm (0.006 – 0.007 in) for 250 models, and 0.15 – 0.21 mm (0.006 – 0.008 in) for 500 models. Renew the pump if the clearance is more than 0.25 mm (0.010 in). Check the clearance between any two rotor peaks in a similar manner. The nominal clearance is 0.15 mm (0.006 in). Renew if worn to 0.20 mm (0.008 in).

3 The side float of the rotors can be checked by laying a straightedge across the pump face, and measuring the gap between it and the rotors. This gap should normally between 0.01 and 0.07 mm (0.0004 – 0.0028 in). If the measurement exceeds 0.12 mm (0.0047 in) the rotors and/or pump body should be renewed.

4 When reassembling the pump, lubricate each component wlth clean engine oil, making sure that it is worked around the rotors. The pump backplate is located by two dowels and is secured by a single screw. The pump body screws, when fitted, also secure the backplate.

13 Oil filter: location and general description

1 The oil filtration arrangement in the XL and XR models is unusual by modern standards and relies on a single small gauze filter. This traps any particles which might otherwise be drawn up into the pump, causing damage to it and the moving parts of the engine and gearbox. It must be appreciated that any debris which is smaller than the holes in the gauze can pass freely into the lubrication system, and whilst this will not normally cause problems, any delay in changing the engine oil can lead to accelerated rates of engine wear. It follows that oil changes must be performed promptly at the specified intervals to prolong the life of the engine.

2 It is advisable to remove and clean the gauze screen at regular intervals, and as it is necessary to drain the oil content of the engine unit to do so, it is convenient to clean the screen at each oil change. To gain access to the filter screen it is necessary to remove the right-hand outer casing. The filter resides in a slot at the bottom of the casing and can be withdrawn using a pair of pointed-nose pliers. Clean the element by washing it thoroughly in clean petrol (gasoline), making sure that all traces of contamination are removed. When clean and dry the element can be refitted and new engine oil added.

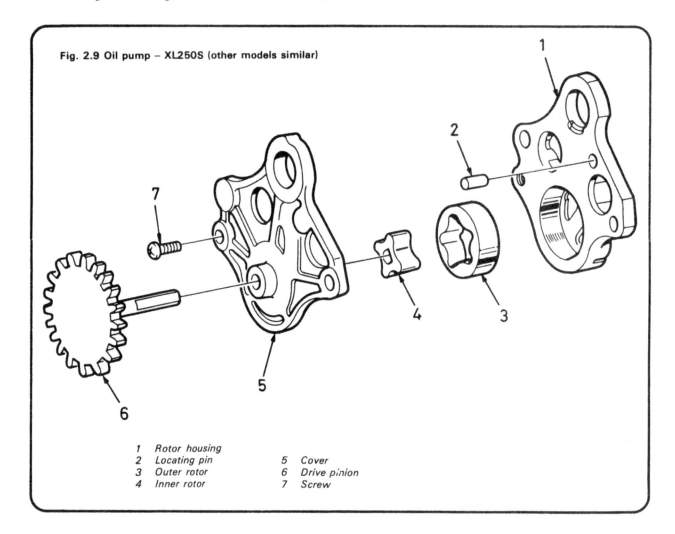

Fig. 2.9 Oil pump – XL250S (other models similar)

1	Rotor housing	5	Cover
2	Locating pin	6	Drive pinion
3	Outer rotor	7	Screw
4	Inner rotor		

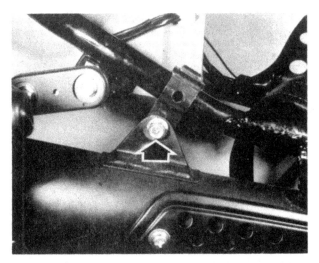

11.5a Remove rear mounting bolt (arrowed) and free bracket

11.5b Single mounting screw (arrowed) passes down through frame

11.5c Silencer can be manoeuvred clear as shown

12.2a Measure clearance between pump body and rotor

12.2b Measure clearance between rotor tips

12.3 Use straight edge to check endfloat

12.4a Fit pinion and spindle through pump cover

12.4b Offer up pump cover to body and rotors

12.4c Pump is secured by single screw...

12.4d ...and located by hollow dowel

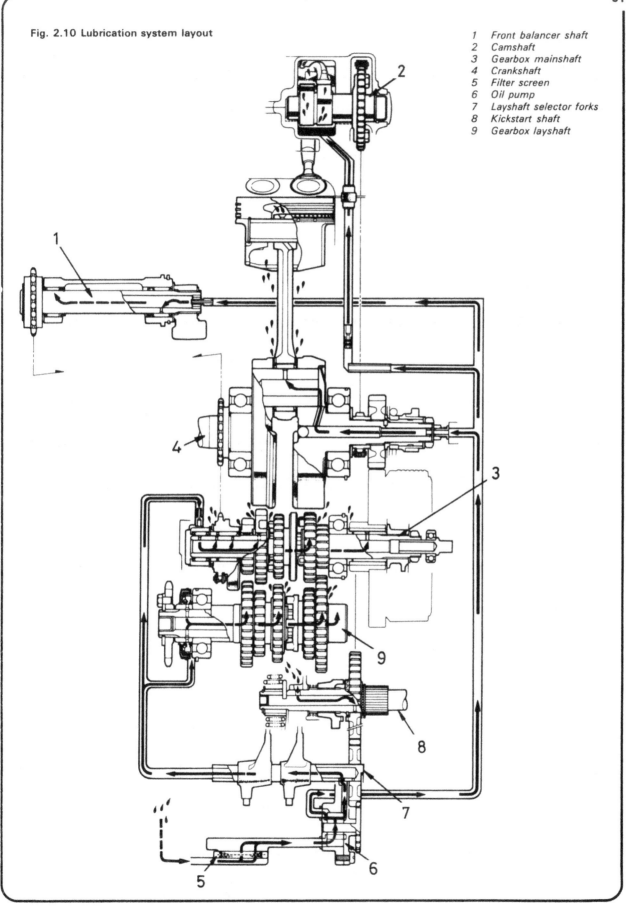

Fig. 2.10 Lubrication system layout

1 Front balancer shaft
2 Camshaft
3 Gearbox mainshaft
4 Crankshaft
5 Filter screen
6 Oil pump
7 Layshaft selector forks
8 Kickstart shaft
9 Gearbox layshaft

14 Fault diagnosis: fuel system and lubrication

Symptom	Cause	Remedy
Excessive fuel consumption	Air cleaner choked or restricted	Clean or renew.
	Fuel leaking from carburettor. Float sticking	Check all unions and gaskets. Float needle seat needs cleaning.
	Float damaged or leaking	Check and renew.
	Float level set too high	Check and adjust (see text).
	Badly worn or distorted carburettor	Replace.
	Jet needle setting too high	Adjust to figure given in Specifications.
	Main jet too large or loose	Fit correct jet or tighten if necessary.
	Carburettor flooding	Check float valve and replace if worn.
Idling speed too high	Throttle stop screw in too far	
	Carburettor top loose	Adjust screw. Tighten top.
	Pilot screw incorrectly adjusted	Refer to relevant paragraph in this Chapter.
	Throttle cable sticking	Disconnect and lubricate or replace.
Engine dies after running for a short while	Blocked air hole in filter cap	Clean.
	Dirt or water in carburettor	Remove and clean out.
General lack of performance	Weak mixture; float needle stuck in seat	Remove float chamber or float and clean.
	Air leak at carburettor joint	Check joint to eliminate leakage, and fit new O-ring
Engine does not respond to throttle	Throttle cable sticking	See above.
	Petrol octane rating too low	Use higher grade (star rating) petrol.
Engine runs hot and is noisy	Lubrication failure	Stop engine immediately and investigate cause. Do not restart until cause is found and rectified.

Note: Incorrect ignition settings can give rise to symptoms similar to some of the above. Check the ignition system in conjunction with the fuel system to eliminate this possibility. (See Chapter 3 for details).

Chapter 3 Ignition system

For modifications and information relating to later models, see Chapter 7

Contents

Specifications

Ignition system
Type .... Capacitor discharge ignition (CDI)

Sparking plug
UK type:
Standard .... NGK DR8ES-L (ND X24ESR-U)
Cold climate (below 5°C, 41°F) .... NGK DR7ES (ND X22ESR-U)
Continuous high-speed use .... NGK DR9ES (ND X27ESR-U)

US type:
Standard .... NGK D8EA (ND X24ES-U)
Cold climate (below 5°C, 41°F) .... NGK D7EA (ND X22ES-U)
Continuous high-speed use .... NGK D9EA (ND X27ES-U)
Gap .... 0.6 – 0.7 mm (0.024 – 0.028 in)

Ignition timing

	250 models	500 models
Initial	12° BTDC @ 1200 ± 100 rpm	10° BTDC @ 2500 ± 250 rpm
Full advance	37 ± 2° BTDC @ 3300 rpm	36° BTDC @ 3500 rpm

1 General description

The models covered by this manual are equipped with a CDI (capacitor discharge ignition) system. The system is powered by a source coil built into the flywheel generator. Power from this coil is fed directly to the CDI unit mounted beneath the frame, where it passes through a diode which converts it to direct (dc). The charge is stored in a capacitor at this stage.

The spark is triggered by the pulser assembly which is mounted on the right-hand end of the crankshaft. As the magnetic rotor passes the pulse coil a small alternating current (ac) pulse is induced. This enters the CDI unit where it is rectified by a second diode. The heart of the CDI unit is a component known as a thyristor. It acts as an electronic switch, which remains off until a small current is applied to its gate terminal. This causes the thyristor to become conductive, and it will remain in this state until any stored charge has discharged through the primary windings of the coil.

The sudden discharge of low-tension energy through the coil's primary windings in turn induces a high tension charge in the secondary coil. It is this which is applied to the centre electrode of the spark plug, where it jumps the air gap to earth (ground), igniting the fuel/air mixture.

As engine speed rises, it becomes necessary for the timing of the ignition spark to be advanced in relation to the crankshaft to allow sufficient time for the combustion of the air/fuel mixture to take place at the optimum position. This function is catered for by a conventional centrifugal automatic timing unit incorporated in the pulser rotor assembly.

2 CDI system: fault diagnosis

1   As no means of adjustment is available, any failure of the system can be traced to the failure of a system component or a simple wiring fault. Of the two possibilities, the latter is by far the most likely. In the event of failure, check the system in a logical fashion, as described below.

2   Remove the sparking plug, giving it a quick visual check, noting any obvious signs of flooding or oiling. Fit the plug into the plug cap and rest it on the cylinder head so that the metal body of the plug is in good contact with the cylinder head metal. The electrode end of the plug should be positioned so that sparking can be checked as the engine is spun over using the kickstart.

3   **Important note**: The energy levels in electronic systems can be very high. On no account should the ignition be switched on whilst the plug or plug cap is being held. Shocks from the HT circuit can be most unpleasant. Secondly, it is vital that the plug is in position and soundly earthed when the system is checked for sparking. The CDI unit can be seriously damaged if the HT circuit becomes isolated.

4   Having observed the above precautions, turn the ignition and engine kill switches to 'On' and kick the engine over. If the system is in good condition a regular, fat blue spark should be evident at the plug electrodes. If the spark appears thin or yellowish, or is non-existent, further investigation will be necessary. Before proceeding further, turn the ignition off and remove the key as a safety measure.

5   Ignition faults can be divided into two categories, namely those where the ignition system has failed completely, and those which are due to a partial failure. The likely faults are listed below, starting with the most probable sources of failure. Work through the list systematically, referring to the subsequent sections for full details of the necessary checks and tests.

Total or partial ignition system failure

1   Loose, corroded or damaged wiring connections, broken or shorted wiring between any of the component parts of the ignition system.
2   Faulty main switch or engine kill switch.
3   Faulty ignition coil.
4   Faulty CDI unit.
5   Faulty alternator.
6   Faulty pulser assembly.

3   CDI system: checking the wiring

1   The wiring should be checked visually, noting any signs of corrosion around the various terminals and connectors. If the fault has developed in wet conditions it follows that water may have entered any of the connectors or switches, causing a short circuit. A temporary cure can be effected by spraying the relevant area with one of the proprietary de-watering aerosols, such as WD40 or similar. A more permanent solution is to dismantle the switch or connector and coat the exposed parts with silicone grease to prevent the ingress of water. The exposed backs of connectors can be sealed off using a silicone rubber sealant.

2   Light corrosion can normally be cured by scraping or sanding the affected area, though in serious cases it may prove necessary to renew the switch or connector affected. Check the wiring for chafing or breakage, particularly where it passes close to part of the frame or its fittings. As a temporary measure,

damaged insulation can be repaired with PVC tape, but the wire concerned should be renewed at the earliest opportunity.

3   Using the wiring diagrams at the end of the manual, check each wire for breakage or short circuits using a multimeter set on the resistance scale or a dry battery and bulb wired as shown in the accompanying illustration. In each case, there should be continuity between the ends of each wire.

4   CDI system: checking the ignition and kill switches

1   The ignition, or main, switch is fitted to the XL versions only. It has four terminals which can be identified as follows. The red lead carries power from the battery to the switch where it is connected to the black lead when the switch is turned on. This is the feed to the horn and lighting systems and can be ignored as far as ignition problems are concerned. The two remaining terminals are the earth, or negative (-) side of the circuit and this terminal is connected to the green lead. The remaining black/white lead runs from the switch, through a multi-pin connector to the CDI unit. When the ignition switch is in the 'Off' position, the black/white lead is connected to the green lead and thus to earth.

2   The purpose of the test is to ensure that the black/white lead and the green lead are connected when the switch is off and isolated when the switch is 'On'. This can be checked using a multimeter or the battery/bulb arrangement described earlier. If the test indicates that the black/white lead is earthed irrespective of the switch position, check that the engine kill switch is set at the 'Run' position. If this fails to affect the result, trace and disconnect the ignition (black/white) and earth (green) leads from the ignition switch. Repeat the test with the switch isolated. If no change is apparent, the switch should be considered faulty and renewed.

3   If the ignition switch works normally when isolated, the fault must lie in the black/white lead between the CDI unit and the ignition and kill switches or in the kill switch itself. On the XL machines the ignition and kill switches perform the same function, each earthing the ignition circuit when set on the 'Off' position, thus unless both are disconnected from earth, the ignition circuit will not operate.

4   The kill switch is fitted to both the XL and the XR versions, and as mentioned above, serves to earth the ignition circuit when switched off. There are two connections to the switch, namely the ignition (black/white) lead and the earth (green) lead. Using the multimeter or battery/bulb arrangement, check that the two terminals are connected when the switch is 'Off' and disconnected when it is turned 'On'. Before condemning the ignition or kill switches, check that the fault has not been caused by contamination with water or dirty contacts.

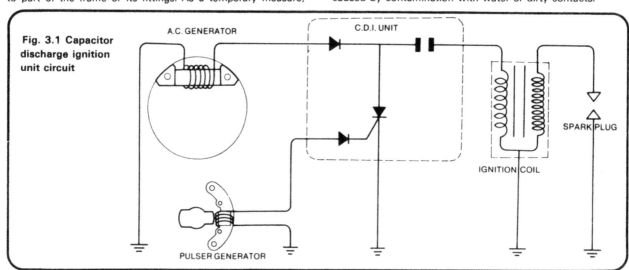

Fig. 3.1 Capacitor discharge ignition unit circuit

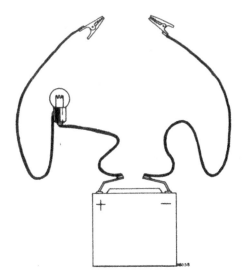

Fig. 3.2 Battery and bulb resistance test

5 Ignition coil: location and testing

1 The ignition coil is a sealed unit, and will normally give long service without need for attention. It is mounted beneath the frame gusseting to the rear of the steering head and is covered in use by the fuel tank. It follows that it will be necessary to remove the tank in order to gain access to the coil.

2 If a weak spark and difficult starting causes the performance of the coil to be suspect, it should, in general, be tested by a Honda service agent or an auto-electrical expert. They will have the necessary appropriate test equipment. It is, however, possible to perform a number of basic tests, using a multimeter with ohms and kilo ohms scales. The primary winding resistance should be checked by connecting one of the meter probe leads to the Lucar terminal and the other earthed against the coil mounting lug. The secondary windings are checked by connecting the probe leads to the high tension lead,

having removed the plug cap, and to the coil mounting lug.

Primary winding resistance	*0.2 – 0.8 ohms*
Secondary winding resistance	*8 – 15 k ohms*

3 Should any of these checks not produce the expected result, the coil should then be taken to a Honda service agent or auto-electrician for a more thorough check. If the coil is found to be faulty, it must be replaced; it is not possible to effect a satisfactory repair.

6 CDI unit: location and testing

1 The CDI unit takes the form of a sealed metal box mounted beneath the fuel tank. In the event of malfunction the unit may be tested in situ after the fuel tank has been removed and the wiring connectors traced and separated. Honda advise against the use of any test meter other than the Sanwa Eletric Tester (Honda part number 07308-0020000) or the Kowa Electric Tester (TH-5H), because they feel that the use of other devices may result in inaccurate readings.

2 Most owners will find that they either do not possess a multimeter, in which case they will probably prefer to have the unit checked by a Honda Service Agent, or own a meter which is not of the specified make or model. In the latter case, a good indication of the unit's condition can be gleaned in spite of small inaccuracies in the readings. If necessary, the CDI unit can be taken to a Honda Service Agent or auto-electrical specialist for confirmation.

3 The unit's six output lead colours and functions are shown in the accompanying diagram, and the readings to be expected from each pair of leads in the table below.

Unit kΩ

+ Probe –	Black/Red	Green	Black/White[1]	Blue/Yellow	Black/White[2]
Black/Red		∞	0.5–10	∞	∞
Green	0.5–10		2–50	∞	∞
Black/White[1]	∞	∞		∞	∞
Blue/Yellow	2–50	0.5–10	2–50		∞
Black/White[2]	∞	∞	∞	∞	

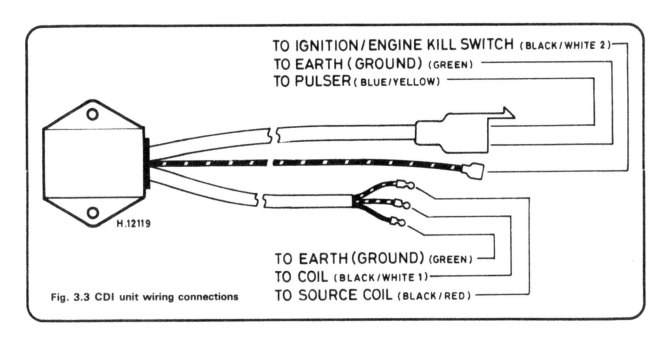

TO IGNITION/ENGINE KILL SWITCH (BLACK/WHITE 2)
TO EARTH (GROUND) (GREEN)
TO PULSER (BLUE/YELLOW)

H.12119

TO EARTH (GROUND) (GREEN)
TO COIL (BLACK/WHITE 1)
TO SOURCE COIL (BLACK/RED)

Fig. 3.3 CDI unit wiring connections

5.1 Ignition coil is mounted on frame below fuel tank

6.1 CDI unit is located below fuel tank

6.3 Connectors; Pickup (A), CDI Unit (B)

is thus wasted. This arrangement is intentional, having no effect on the running of the engine, but making the triggering of the spark easier to contrive. Neither the stator or rotor parts of the unit are adjustable, making timing adjustment impossible and thus regular timing checks unnecessary.

2   In the event that the pulser assembly fails to pulse, it will be seen that the CDI unit will not be triggered, and thus sparks at the plug electrodes will be noticeably lacking. Generally speaking, the assembly is not likely to cause problems. It is theoretically possible for the rotor to become demagnetised to the point where sparking becomes unreliable, but in practice the tiny current required to trigger the CDI unit is invariably produced. A short or open circuit in the pulser coil is more likely, and can be checked by measuring the coil resistance.

3   Trace the pulser leads back to the connector beneath the fuel tank. Separate the connector and measure the resistance between the green and the blue/yellow wires. In good condition, the pulser coil should show a resistance of 20-60 ohms. A reading of infinity or zero resistance will be indicative of an open or short circuit respectively, and will necessitate renewal of the coil. Should renewal be required, take the rotor assembly to ensure that the new coil unit has the same matching mark. An unmatched pulser coil and rotor can cause poor engine performance.

7   Alternator source coil: testing

1   If the alternator source coil is suspected of malfunction, it can be checked by measuring its resistance. Trace and disconnect the Black/red lead at the CDI unit connector. Set the multimeter to the resistance (ohms) scale and connect one probe lead to the Black/red lead and the other to earth (ground). The meter should indicate a resistance of 200-500 ohms if the source coil is in good condition. An infinite resistance or zero resistance indicates an open or short circuit respectively, and will require the renewal of the source coil.

8   Pulser assembly: testing

1   The pulser assembly consists of a small coil unit, mounted on the inside of the right-hand outer casing, and a magnetic rotor or reluctor which is attached to the end of the crankshaft. As the crankshaft rotates, the rotor tip passes the coil pole inducing a small trigger current which is then fed to the thyristor gate in the CDI unit. It will be appreciated that, in a four-stroke application, this system produces two sparks per complete engine cycle, one of which occurs during the exhaust stroke and

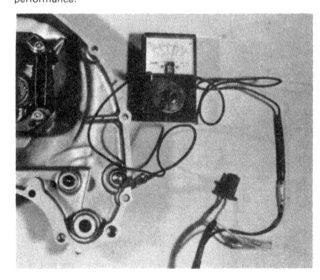

7.1 Multimeter can be used to check source coil resistance

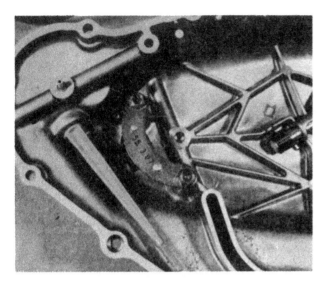

8.1a Pulser coil is mounted inside outer casing

8.1b Magnetic poles trigger the ignition pulse (arrowed)

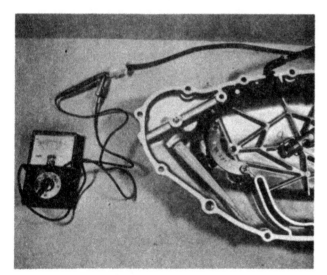

8.3 Pulser resistance is checked as shown

9 Ignition timing: checking

1   As a result of the adoption of electronic ignition, periodic checking of the ignition timing should not be necessary, and indeed, no provision is given for adjustment. Checking the timing is worthwhile, however, if a new pulser or reluctor is fitted, and it is also a useful means of checking the function of the mechanical ATU if its performance is suspect.

2   The ignition timing can be checked only whilst the engine is running using a stroboscopic lamp and thus a suitable lamp will be required. Furthermore some means of determining the engine speed is required, and because a tachometer is not fitted as standard, a test tachometer must be used. The ignition timing is checked by reference to timing marks scribed on the periphery of the generator rotor and a fixed index mark on the generator cover. To gain access to the marks remove the upper inspection cap from the cover.

3   To check the ignition timing connect the strocoscope to the machine's ignition system high-tension or low-tension circuit as directed by the lamp's manufacturer. Where necessary connect the test tachometer. Start the engine and aim the lamp at the generator rotor. If the ignition timing is correct the 'F' mark will

align with the index mark at the engine speed given for the initial (retarded) timing in the table below. Raise the speed of the engine until the full advance speed given is obtained. At this speed the index mark should align with the two parallel advance marks scribed on the generator rotor. The transition in the timing from retarded to full advance should take place smoothly in proportion to the speed. If it can be seen that the movement is erratic or if full advance cannot be reached then there is some indication that the advance mechanism is not functioning correctly. Refer to the following Section for details of ATU inspection.

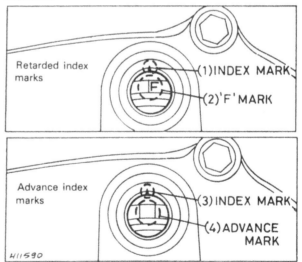

Fig. 3.4 Ignition timing marks

10 Automatic timing unit (ATU): location, function and testing

1   Despite the adoption of the CDI system in place of the earlier coil-and-contact breaker systems, Honda have elected to retain a mechanical centrifugal advance system in the form of an automatic timing unit. This device is necessary to cause the ignition spark to take place progressively earlier as the engine speed increases, so that the fuel/air mixture is given time to burn in a manner which will produce the maximum amount of useful work.

2   The automatic timing unit consists of a baseplate fitted with a central spindle, by which it is retained to the crankshaft end through a taper and securing bolt, and two pivot pins arranged at the outer edge of the plate at 180° to each other. Two bobweights are fitted over the pivot pins and anchored by a pair of small tension springs. Tangs on the weights engage in the ignition pulser rotor (reluctor) which is mounted concentrically over the centre spindle.

3   As the engine runs, the ATU is spun on the end of the crankshaft. This results in the weights being flung outwards by centrifugal force, and in doing so, they cause the rotor to move in relation to the crankshaft. The weights are controlled by the light tension springs, and these pull the weights back as the engine slows down.

4   To gain access to the ATU, it is necessary to remove the right-hand outer cover after draining the engine oil. The rotor and ATU unit is secured by a single nut. It may prove necessary to lock the crankshaft whilst the nut is removed. This can best be accomplished by selecting top gear and applying the rear brake. Remove the nut and slide the ATU off the crankshaft end. It is not necessary to disturb the spring-loaded oil feed quill.

5   Examine each component for wear, checking that none of the anchor and pivot pins have become loose in the backplate. It may be possible to secure loose pins by re-riveting them, but often the whole assembly will require renewal. Similarly, only a small amount of wear can be tolerated before sloppy operation allows the ignition setting to wander.

6   There is little point in attempting to dismantle the unit since the component parts are not available separately. If a new unit is required it is essential that it matches the pickup or pulser coil and that the correct unit for any given model year is fitted. To this end, the engine and frame number should be given in full.

7   The action of the ATU during operation can be checked visually using a stroboscopic lamp as discussed in the preceding Section.

11 Sparking plug: checking and setting the gap

1   A range of plug types are available for the XL and XR models, the type chosen depending on the application and operating conditions. These are shown in the Specifications at the beginning of this Chapter. In general, the Australian, South African and General Market models have similar plug applications to the US versions, whilst the UK, European and Canadian models can be grouped in a separate category.

2   The correct electrode gap for all models is 0.6 – 0.7 mm (0.024 – 0.028 in). The gap can be assessed using feeler gauges. If necessary, alter the gap by moving the outer electrode, preferably using a proper electrode tool. **Never** bend the centre electrode, otherwise the porcelain insulator will crack, and may cause damage to the engine if particles break away whilst the engine is running.

4   After some experience the sparking plug electrodes can be used as a reliable guide to engine operating conditions. See accompanying photographs.

5   It is advisable to carry a new spare sparking plug on the machine, having first set the electrodes to the correct gap. Whilst sparking plugs do not fail often, a new replacement is well worth having if a breakdown does occur.

6   Never overtighten a sparking plug otherwise there is risk of stripping the threads from the cylinder head, especially as it is cast in light alloy. A stripped thread can be repaired without having to scrap the cylinder head by using a 'Helicoil' thread insert. This is a low-cost service, operated by a number of dealers.

7   Before replacing a sparking plug into the cylinder head coat the threads sparingly with a graphited grease to aid future removal. Use the correct size spanner when tightening the plug otherwise the spanner may slip and damage the ceramic insulator. The plug should be tightened sufficiently to seat firmly on the sealing washer, and no more.

12 Fault diagnosis: ignition system

Symptom	Cause	Remedy
Engine will not start	Spark at plug weak or non-existent	Fouled or faulty plug – renew Check ignition wiring and connections Plug cap, HT lead or coil faulty – renew CDI unit or pulser faulty – renew
Engine starts but runs erratically	Intermittent or weak spark	Check system as described above
As above, but engine tends to kick back when started. Engine runs hot	Automatic timing unit jammed open	Examine and clean or renew as required
Engine starts but runs sluggishly. Tends to overheat	Automatic timing unit jammed closed	See above
Ignition system fails when used in rain	Water shorting ignition or engine kill switch	Use water displacing spray such as WD40 or Contact to locate fault and effect temporary cure. Waterproof component at a later date

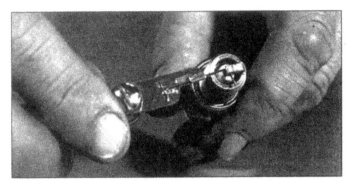

Electrode gap check - use a wire type gauge for best results

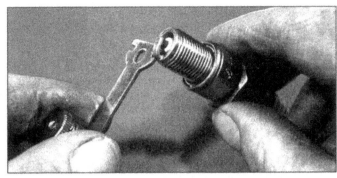

Electrode gap adjustment - bend the side electrode using the correct tool

Normal condition - A brown, tan or grey firing end indicates that the engine is in good condition and that the plug type is correct

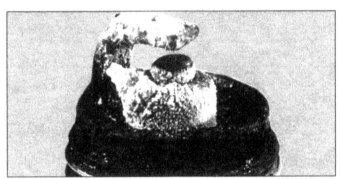

Ash deposits - Light brown deposits encrusted on the electrodes and insulator, leading to misfire and hesitation. Caused by excessive amounts of oil in the combustion chamber or poor quality fuel/oil

Carbon fouling - Dry, black sooty deposits leading to misfire and weak spark. Caused by an over-rich fuel/air mixture, faulty choke operation or blocked air filter

Oil fouling - Wet oily deposits leading to misfire and weak spark. Caused by oil leakage past piston rings or valve guides (4-stroke engine), or excess lubricant (2-stroke engine)

Overheating - A blistered white insulator and glazed electrodes. Caused by ignition system fault, incorrect fuel, or cooling system fault

Worn plug - Worn electrodes will cause poor starting in damp or cold weather and will also waste fuel

Chapter 4 Frame and forks

For modifications and information relating to later models, see Chapter 7

Contents

Specifications

	XL250S	XR250	XL500S	XR500
Frame				
Type	Welded tubular steel with welded steel spine			
Front forks				
Type	Oil damped telescopic			
Travel	204 mm	224 mm	204 mm	224 mm
	(8.0 in)	(8.8 in)	(8.0 in)	(8.8 in)
Oil capacity per leg	185–195 cc	200–205 cc	185–195 cc	200–205 cc
Oil grade (all models)	Automatic transmission fluid (ATF)			
Fork spring free length:				
Main spring	510.9 mm	562.4 mm	481.9 mm	562.4 mm
	(19.76 in)	(22.1 in)	(18.97 in)	(22.1 in)
Service limit	483.5 mm	551.0 mm	477.7 mm	552.7 mm
	(19.04 in)	(21.7 in)	(18.81 in)	(21.8 in)
Secondary spring	N/A	—	69.8 mm	—
			(2.75 in)	
Service limit	39.0 mm	—	68.0 mm	—
	(1.54 in)		(2.68 in)	
Rear suspension				
Type	Pivoted rear fork (swinging arm)			
Travel	178 mm	198 mm	178 mm	198 mm
	(7.0 in)	(7.8 in)	(7.0 in)	(7.8 in)
Rear suspension units	Coil spring, 5-way adjustable preload, gas/oil damper units			
Spring free length	326.9 mm	345.8 mm	325.5 mm	345.8 mm
	(12.87 in)	(13.6 in)	(12.8 in)	(13.6 in)
Service limit	320.5 mm	339.0 mm	321.8 mm	332.0 mm
	(12.61 in)	(13.3 in)	(12.7 in)	(13.1 in)
Swinging arm bush clearance	0.2–0.3 mm	0.2–0.3 mm	0.040–0.125 mm	0.040–0.125 mm
	(0.008–0.012 in)	(0.008–0.012 in)	(0.0016–0.0049 in)	(0.0016–0.0049 in)
Service limit	0.8 mm	0.8 mm	0.32 mm	0.32 mm
	(0.032 in)	(0.032 in)	(0.013 in)	(0.013 in)

1 General description

The Honda XL and XR models featured in this manual each employ a welded tubular steel frame built around a fabricated sheet steel spine. No frame cradle is employed, the engine acting as a stressed member of the overall assembly. Front suspension is of the conventional oil-damped telescopic fork variety, whilst rear suspension is provided by a pivoted rear fork controlled by oil/gas-damped coil spring suspension units.

2 Front forks: removal – general

1   It is unlikely that the forks will require removal from the frame unless the fork seals are leaking or accident damage has been sustained. In the event that the latter has occurred, it should be noted that the frame may also have become bent, and whilst this may not be obvious when checked visually, could prove to be potentially dangerous.
2   If attention to the fork legs only is required, it is unnecessary to detach the complete assembly, the legs being easily removed individually.
3   If attention to the steering head assembly is required it is possible to remove the lower yoke with the fork legs still in place, if desired. It should be noted, however, that this procedure is hampered by the unwieldy nature of the assembly, and it is recommended that the fork legs be removed prior to dismantling the steering head and fork yokes.
4   Before dismantling work can begin it will be necessary to contrive some means of supporting the machine with the front wheel clear of the ground, as no centre stand is fitted. The best method is to use a stout wooden crate or similar placed beneath the engine skid plate. In the interests of safety and the continued integrity of the machine's paint finish use ropes or tie-down straps to support it in an upright position.
5   Release the speedometer drive cable by unscrewing the screw which retains it in the brake plate. Slacken the brake cable adjuster and disconnect the cable at the lower end. Straighten and remove the split pin (where fitted) which passes through the castellated wheel spindle nut. Models fitted with a domed spindle nut do not have a split pin. Slacken the nut, then release the pinch bolt(s) which clamp the opposite end of the spindle. Support the wheel while the spindle is displaced, then remove the wheel and place it to one side.
6   The front mudguard is bolted to the underside of the bottom fork yoke and need not be disturbed when removing the fork legs. If, however, the steering head assembly is to be dismantled, it will be necessary to release it by removing its retaining bolts. Take care that the rubber mounting washers are not lost during removal.

3 Front forks: removing the fork legs from the yoke

1   It is not necessary to remove the complete headstock assembly if attention to the fork legs alone is required. The instructions in Section 2 of this Chapter should be followed, then proceed as described below.
2   Slacken the upper and lower pinch bolts which retain each fork leg. It should now be possible to pull and twist the fork legs downward to disengage them from the yokes. If necessary, a wooden drift can be used to knock the legs downward and clear of the yokes.

4 Steering head assembly: dismantling

XL models
1   To gain access to the steering head components it will be necessary to remove the various ancillary parts attached to it. Start by removing the front wheel and fork legs as described in Sections 2 and 3. Remove the seat by releasing its two securing bolts and, where appropriate, the retaining strap. Turn the fuel tap off and prise off the fuel feed pipe. Release the single tank fixing bolt and remove the fuel tank, placing it in a safe place away from any possible fire risk.
2   Remove the headlamp retaining screws and lift the assembly clear of the headlamp shell. Disconnect the various wiring connectors inside the shell and push them out through the holes at the rear. The shell can now be removed after unscrewing the reflectors or bolts which retain it to its brackets.
3   Release the speedometer cable by unscrewing its retaining ring. Trace and disconnect the instrument panel wiring where this has not already taken place. Three groups of wires run from the instrument panel, carrying the ignition switch, warning lamp and illumination wiring. The panel can be removed separately by removing the two domed nuts which secure it to the bracket assembly. It is preferable, however, to remove the panel and bracket as an assembly by removing the two bolts which retain the latter to the top yoke.
4   The handlebar assembly should be removed, a process which, in theory, demands the removal of the controls, cables and indicator lamps. In practice, much unnecessary work can be avoided by simply releasing the handlebar clamps and resting the assembly across the frame spine, clear of the top yoke. The two clamps are retained by a total of four bolts.
5   Slacken and remove the large chromium-plate steering stem nut, using a box spanner or socket to prevent damage to the finish. The space between the handlebar mountings precludes the use of adjustable spanners or an open-ended spanner, which would in any case tend to mar the chromium plating.
6   The top yoke can now be removed by pulling it upwards, until it disengages from the steering stem. It will probably be necessary to tap the yoke upwards to free it, and this must be done with a soft-faced mallet to preclude damage to the yoke. Alternatively, a block of wood can be used to spread the impact of hammer blows. Note that great care must be taken when striking the yoke. The effects of a subsequent fracture due to indiscriminate hammering would be disastrous at the very least.
7   The forks and lower yoke can be removed downwards as a single unit. Note that on all models, the steel balls from the steering head lower bearing race will drop free as the assembly is withdrawn. Those with previous experience of these bearings may well know of their remarkable property of bouncing into the darkest recesses of the workshop, so unless new bearings are to be fitted as a matter of course, some clean rag should be arranged below the headstock to contain the balls as they drop free. If this precaution is not taken it is worthwhile knowing that the subsequent search should produce 18 elusive $\frac{1}{4}$ in steel balls from the upper and lower races, a total of 36 in all.
8   To prevent loss, it is a good idea to tape a plastic bag around the lower yoke (not to the steering head lug) so that any errant balls are contained. Should any mishap occur despite these precautions, a magnet would be most useful in locating lost balls. Using a C-spanner, slacken the slotted steering stem nut whilst supporting the lower yoke. When the nut has been removed the assembly can be lowered clear of the steering head. The balls in the lower bearing race will drop free, but those in the upper race will probably remain in position. These and any others clinging to the underside of the steering head should be collected and placed in a suitable container to await reassembly.

XR models
9   The procedure for dismantling the XR versions is rather simpler than that described for the XL models, due to the relative simplicity of the various fittings. The headlamp and front number plate are removed as a unit after releasing the two upper mounting bolts and the lower mounting bolt. Disconnect the headlamp leads at the appropriate connectors located behind the unit.
10 Slacken the speedometer cable knurled nut and disconnect the cable. Remove the speedometer retaining nuts and lift the instrument clear of the top yokes. The rest of the dismantling procedure is the same as that described for the XL models.

3.2a Slacken the lower yoke pinch bolt(s)...

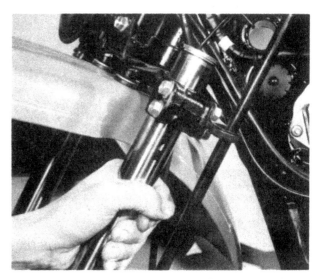

3.2b ...and upper yoke pinch bolts. Withdraw fork leg

5   Steering head bearings: examination and renovation

1   Before commencing reassembly of the forks, examine the steering head races. The ball bearing tracks of the respective cup and cone bearings should be polished and free from indentations, cracks or pitting. If signs of wear are evident, the cups and cones must be renewed. In order for the straight line steering on any motorcycle to be consistently good, the steering head bearings must be absolutely perfect. Even the smallest amount of wear on the cups and cones may cause steering wobble at high speeds and judder during heavy front wheel braking. The cups and cones are an interference fit on their respective seatings and can be tapped from position with a suitable drift.

2   Ball bearings are relatively cheap. If the originals are marked or discoloured they **must** be renewed. To hold the steel balls in place during reassembly of the fork yokes, pack the bearings with grease. The upper and lower races each contain 18 $\frac{1}{4}$ in steel balls. Although a small gap will remain when the balls have been fitted, on no account must an extra ball be inserted, as the gap is intended to prevent the balls from skidding against each other and wearing quickly.

6   Front yokes: examination

1   To check the top yoke for accident damage, push the fork stanchions through the bottom yoke and fit the top yoke. If it lines up, it can be assumed the yokes are not bent. Both must also be checked for cracks. If they are damaged or cracked, fit new replacements.

7   Fork legs: dismantling

1   The front forks legs of all models are of essentially similar design, although there are detail differences in the component parts. The method of dismantling is the same for each type. The fork top bolt has an internal hexagon which is covered by a plastic cap. The top bolt can be removed using a 14 mm Allen key, or failing this by using a pair of nuts locked together on a suitable bolt. The nuts are fitted into the top bolt recess and a conventional spanner used to slacken it.

2   Remove the top bolts and invert the fork legs over a suitable container, propping them until the oil content has drained. The spring(s) and spacers can now be removed from each leg. The XL models employ two fork springs in each leg,

the longer of the two uppermost. A spacer is fitted between the springs, and in the case of the XL500S, a second spacer fits between the top bolt and the upper spring. The XR versions have single springs with a spacer fitted at the top.

3   To separate the fork stanchion and lower leg (fork tube and slider) it will be necessary to slacken and remove the Allen bolt which passes up through the base of the latter and into the damper piston. This is normally a straightforward operation, but occasionally the damper piston will turn in the lower leg making removal difficult. If this situation arises, the manufacturer recommends that the fork spring(s) and top bolt are temporarily refitted to apply pressure to the damper piston. If this method fails, obtain a length of one inch diameter wooden dowel and form a taper on one end. Introduce the tapered end down through the stanchion so that it contacts the recessed head of the damper piston. An assistant should apply pressure to the damper head whilst the bolt is slackened. Note that this operation will be greatly facilitated if the lower leg is clamped in a vice, fitted with soft jaws or rag to protect the soft alloy.

4   When the damper bolt has been released, pull off the dust seal around the top of the lower leg and then pull the stanchion free. The damper seat can be tipped out of the lower leg and the damper piston and rebound spring displaced from the stanchion in a similar manner.

7.1a Remove plastic cap to reveal recessed hexagon

7.1b Bolt can be removed using nut and socket

7.2a Release bolt and washer...

7.2b ...and withdraw fork spring(s)

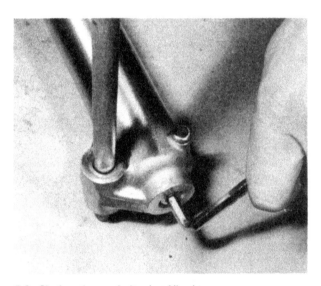

7.3a Slacken damper bolt using Allen key

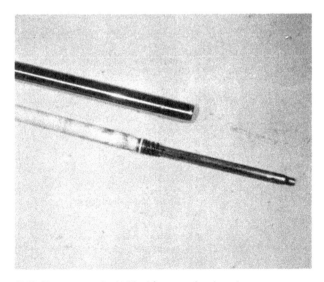

7.3b Damper can be held with tapered rod as shown

7.4a Displace dust seal from lower leg...

7.4b ...and withdraw stanchion assembly

7.4c Damper can be shaken out of stanchion

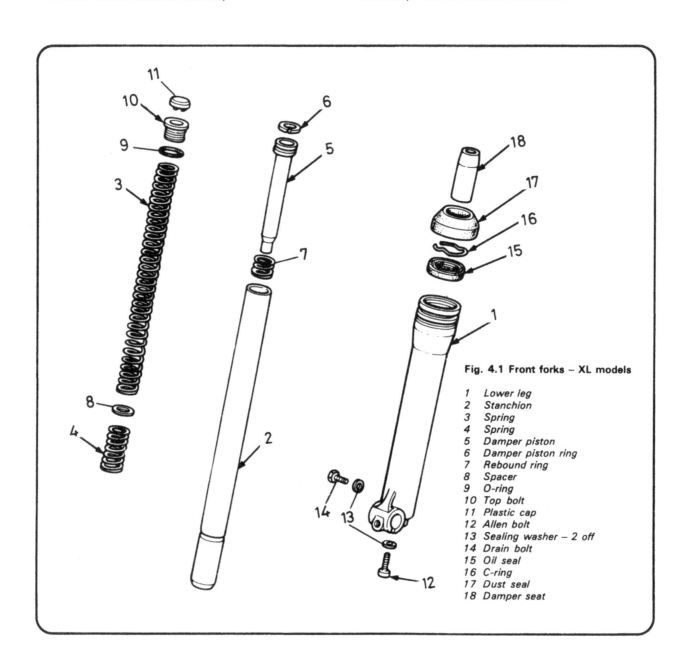

Fig. 4.1 Front forks – XL models

1   Lower leg
2   Stanchion
3   Spring
4   Spring
5   Damper piston
6   Damper piston ring
7   Rebound ring
8   Spacer
9   O-ring
10 Top bolt
11 Plastic cap
12 Allen bolt
13 Sealing washer – 2 off
14 Drain bolt
15 Oil seal
16 C-ring
17 Dust seal
18 Damper seat

Fig. 4.2 Front forks – XR models

1   Lower leg
2   Stanchion
3   Damper pin
4   Spring
5   Damper seat
6   Rebound spring
7   O-ring
8   Top bolt
9   Plastic cap
10 Clamp
11 Screw
12 Gaiter
13 Dust seal
14 C-ring
15 Oil seal
16 Backup ring
17 Allen bolt
18 Sealing washer
19 Stud – 4 off
20 Drain bolt
21 Sealing washer
22 Spindle clamp
23 Nut – 4 off
24 Damper piston ring

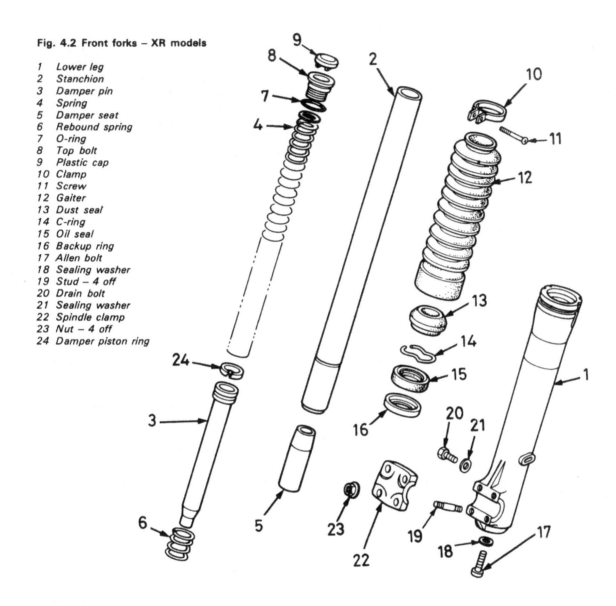

8   Front forks: examination and renovation

1   The parts most likely to wear over an extended period of service are the internal surfaces of the lower leg and the outer surfaces of the fork stanchion or tube. If there is excessive play between these two parts they must be replaced as a complete unit. Check the fork tube for scoring over the length which enters the oil seal. Bad scoring here will damage the oil seal and lead to fluid leakage.

2   It is advisable to renew the oil seals when the forks are dismantled even if they appear to be in good condition. This will save a strip-down of the forks at a later date if oil leakage occurs. The oil seal in the top of each lower leg is retained by an internal C-ring which can be prised out of position with a small screwdriver. With the exception of the XL250S model the seal is supported by a backing ring.

3   Check that the dust excluder rubbers are not split or worn where they bear on the fork tube. A worn excluder will allow the ingress of dust and water which will damage the oil seal and eventually cause wear of the fork tube. The XR models are equipped as standard with full length fork gaiters. These substantially reduce the likelihood of damage due to abrasive dust becoming trapped against the stanchion and scoring it. It follows that the gaiters should be intact and must be refitted correctly during reassembly. Owners of the XL models might consider the fitting of full gaiters for the above reasons, especially where the machine is used predominantly for off-road purposes.

4   It is not generally possible to straighten forks which have been badly damaged in an accident, particularly when the correct jigs are not available. It is always best to err on the side of safety and fit new ones, especially since there is no easy means to detect whether the forks have been over stressed or metal fatigued. Fork stanchions (tubes) can be checked, after removal from the lower legs, by rolling them on a dead flat surface. Any misalignment will be immediately obvious.

5   The fork springs will take a permanent set after considerable usage and will need renewal if the fork action becomes spongy. The length of the fork springs should be checked against the figures given in the Specifications Section. Note that in the case of XL models, two compression springs are

employed to give varying spring characteristics. A further small spring is employed to cushion rebound. This arrangement provides adequate suspension for road work and for off-road use, whilst the single spring XR type is designed purely for off-road applications.

6   The damping action of the forks is governed by the viscosity of the oil in the fork legs, and the recommended types and capacities are given in the Specifications at the beginning of this Chapter. Note that when the fork is being topped up or the damping oil changed, slightly less oil will be needed. It is recommended that a piece of wire is used as a dipstick in these cases, and the oil level measured from the top of the fork stanchion. Given that the initial oil capacity is correct it will be possible to translate this to a known oil level, and subsequent refills and experiments in oil level changes can be made on this basis.

7   It is possible to increase or reduce the damping effect by using a different oil grade, and XR owners in particular may wish to experiment a little to find a grade which suits their particular application. It is advisable to consult a Honda Service Agent who will be able to suggest which oils may be used.

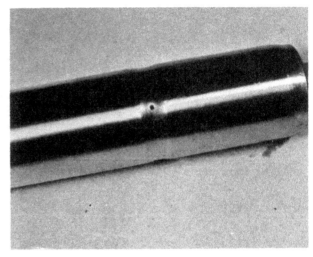

8.1a Check stanchion for scratches – clear oil hole

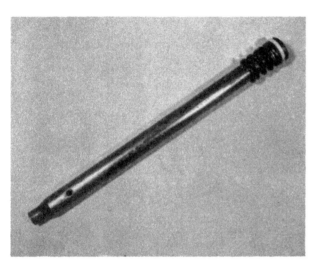

8.1b Check condition of damper assembly

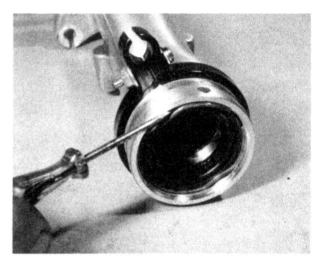

8.2 Prise out the spring retainer to free oil seal

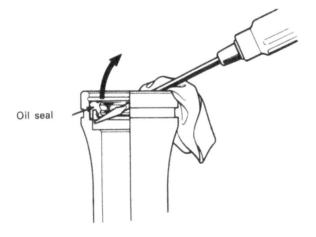

Oil seal

Fig. 4.3 Method of front fork oil seal removal

9   Steering head assembly: reassembly

1   The steering head assembly should be rebuilt, using new bearings and races as required, in the reverse order of the dismantling sequence described in Section 4. The bearing races should be liberally coated with grease. This will provide the necessary lubrication and will make it possible to stick the $18\frac{1}{4}$ in steel balls in position on each race.

2   Carefully position the lower yoke and steering stem, placing the upper cone over the protruding steering stem and securing the assembly by fitting the slotted adjuster nut finger-tight. It is essential that a proper C-spanner is used to tighten the adjuster, because this adjustment must be precise.

3   Honda recommend that the nut is tightened until significant resistance is felt, and then backed off by $\frac{1}{8}$ turn. The object of this is to set the nut so that no discernible free play is allowed, but this must not be at the expense of applying a preload to the bearings. Should this occur, rapid wear, insensitive steering and possibly cracked races or balls could result.

4   Place the top yoke over the steering stem and fit the steering stem top nut. Do not tighten it at this stage, because

there must be some room for alignment of the yokes when the fork legs are refitted. Once these are in place, and the yokes in line, secure the top nut and complete the installation of the ancillary components.

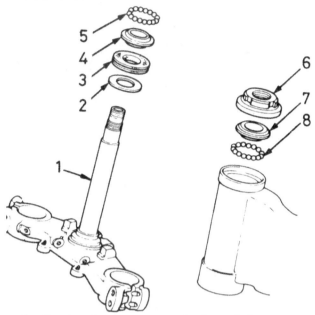

1	Steering stem	5	Lower ball race
2	Washer	6	Slotted nut
3	Dust seal	7	Upper bearing cone
4	Lower bearing cone	8	Upper ball race

Fig. 4.4 Steering head assembly – XL models

10 Front fork legs: reassembly

1   All of the fork leg components should be completely clean and free from dust or oil prior to reassembly. Remember that the forks are in constant motion in use, and any abrasive particles will quickly wear away the surface against which they are trapped. If new seals are required they should be fitted at this stage, noting the backing ring which is fitted first (all models except XL250S). A large diameter socket can be used to drive the new seal into position. Lubricate the seal lip with grease.

2   Fit the damper assembly and rebound spring into the stanchion and drop the damper seat into the lower leg. Feed the stanchion into the lower leg, taking care not to damage the seal. Check that the damper bolt threads are clean and dry, then coat them with Locktite. Fit the damper bolt and tighten to the appropriate torque setting.

Fork damper bolt torque setting

XL250S	*0.8 – 1.2 kgf m (6 – 9 lbf ft)*
All other models	*1.5 – 2.5 kgf m (11 – 18 lbf ft)*

3   Check that the wire retaining clip which secures the oil seal is in place, then fit the dust seal to the top of each lower leg. Fit the fork spring(s) and spacer(s). Fill each fork leg with the appropriate amount and grade of damping oil.

Fork oil capacity

Model	Quantity
XL250S and XL500S	*185 – 195 cc*
	6.51 – 6.86 Imp fl oz
	6.25 – 6.59 US fl oz
XR250 and XR500	*200 – 205 cc*
	7.04 – 7.22 Imp fl oz
	6.76 – 6.93 US fl oz

4   On the XR models fit and locate the fork gaiter, placing the retaining clip loosely around the collar at the top. Fit the fork top bolts, tightening them to 1.5 – 3.0 kgf m (11 – 22 lbf ft). The stanchion may be clamped in a vice with soft jaws during the tightening operation, or better still, slid into the lower yoke. If the lower yoke pinch bolts are tightened temporarily this will serve to hold the stanchion whilst the top bolts are secured.

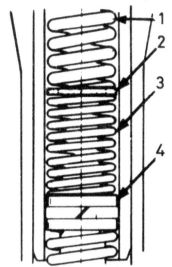

Fig. 4.5 Forks spring assembly – XL models

1	Main fork spring	3	Secondary spring
2	Spacer	4	Damper rod

11 Front forks: replacement

1   Replace the front forks by following in reverse the dismantling procedures described in Section 3 of this Chapter. Before fully tightening the front wheel spindle clamps and the fork yoke pinch bolts, bounce the forks several times to ensure they work freely and are clamped in their original settings. Complete the final tightening from the wheel spindle clamp upwards.

2   Do not forget to add the recommended quantity of fork damping oil to each leg before the bolts in the top of each fork leg are replaced. Check that the drain plugs have been re-inserted and tightened before the oil is added.

3   If the fork stanchions prove difficult to re-locate through the fork yokes, make sure their outer surfaces are clean and polished so that they will slide more easily. It is often advantageous to use a screwdriver blade to open up the clamps as the stanchions are pushed upward into position.

4   As the fork legs are fed through the yokes they will correct any slight misalignment between them should the steering head have been dismantled. Where appropriate, tighten the steering stem top nut and complete the fitting of the various ancillary components. On XR models slide the upper end of the fork gaiter into position and secure it with its retaining clip.

5   Before the machine is used on the road, check the adjustment of the steering head bearings. If they are too slack, judder will occur. There should be no detectable play in the head races when the handlebars are pulled and pushed, with the front brake applied hard.

6   Overtight head races are equally undesirable. It is possible to unwittingly apply a loading of several tons on the head bearings by overtightening, even though the handlebars appear to turn quite freely. Overtight bearings will cause the machine to roll at low speeds and give generally imprecise handling with a tendency to weave. Adjustment is correct if there is no perceptible play in the bearings and the handlebars will swing to full lock in either direction, when the machine is supported with the front wheel clear of the ground. Only a slight tap should cause the handlebars to swing.

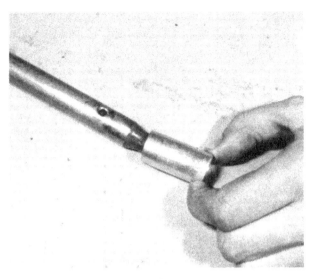

10.2a Damper seat locates as shown.

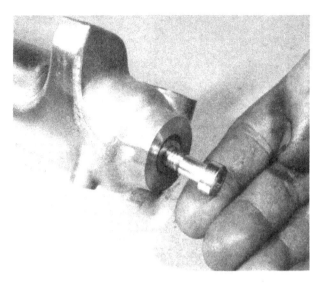

10.2b Use Loctite or similar on damper bolt threads

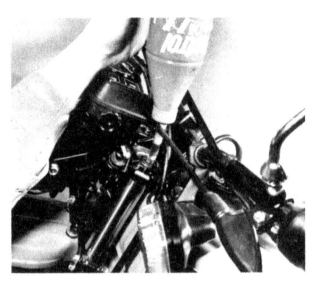

10.3 Top up fork legs with recommended quantity of oil

11.1a Install fork leg and secure clamp bolts

11.1b Tighten top bolt using nut and socket

11.1c Top bolt should be flush with top yoke

12 Frame: examination and renovation

1   The frame is unlikely to require attention unless accident damage has occurred. In some cases, replacement of the frame is the only satisfactory course of action if it is badly out of alignment. Only a few frame repair specialists have the jigs and mandrels necessary for resetting the frame to the required standard of accuracy and even then there is no easy means of assessing to what extent the frame may have been over-stressed.

2   After the machine has covered a considerable mileage, it is advisable to examine the frame closely for signs of cracking or splitting at the welded joints. Rust can also cause weakness at these joints. Minor damage can be repaired by welding or brazing, depending on the extent and nature of the damage.

3   Remember that a frame which is out of alignment will cause handling problems and may even promote 'speed wobbles'. If misalignment is suspected, as the result of an accident, it will be necessary to strip the machine completely so that the frame can be checked and, if necessary, renewed.

13 Rear suspension unit: removal and examination

1   The models featured in this manual are equipped with coil spring suspension units using oil/gas-filled damper assemblies. The pressurised gas provides a supplementary and progressive springing medium, in addition to the external coil springs. The units are mounted at a steep angle to provide maximum rear wheel travel.

2   It is best to remove and attend to one unit at a time, as this will alllow the machine to be supported by the remaining unit. Alternatively, place the machine on a wooden crate or some similar support so that the rear wheel is raised clear of the ground and the weight taken from the suspension units.

3   The units are retained by a single mounting bolt at each end, the upper bolt also retaining the seat strap on the XL models. With the bolts removed, the units can be pulled clear of the frame.

4   The units can be dismantled after the spring has been compressed sufficiently for the lower seat to be disengaged and displaced, a slot providing clearance to enable it to be slid around the damper rod. Ideally, this should be done by compressing the unit with the appropriate spring compressor. It was found in practice that one person could compress the spring adequately by hand, whilst an assistant removed the lower spring seat. If the latter method is chosen, great care must be taken to avoid injury or damage as a result of the assembly slipping under tension.

5   Measure the free length of the spring, and renew it if it below the service limit specified. The damper unit is sealed and cannot be dismantled. Its operation can be checked to some extent by compressing it and then releasing it. The unit should show significant damping effect on rebound (extension) but much less under compression. Any signs of leakage will necessitate renewal of the damper units as a pair.

6   If the suspension units appear to be in need of renewal, thought should be given to fitting an improved proprietary replacement pair. These may prove to be more expensive, but usually provide better control and durability. Most competition accessory shops will be able to advise and recommend the best make and type to use for any given application.

7   Reassembly of the units is a straightforward reversal of the dismantling sequence. When fitting the spring adjuster, set it at its softest position to ease spring compression. When the units have been refitted set the spring adjusters at the required setting, ensuring that the position of each adjuster matches the other.

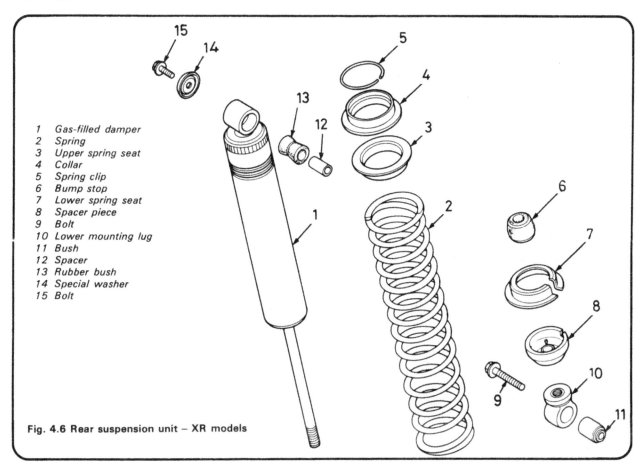

1   Gas-filled damper
2   Spring
3   Upper spring seat
4   Collar
5   Spring clip
6   Bump stop
7   Lower spring seat
8   Spacer piece
9   Bolt
10 Lower mounting lug
11 Bush
12 Spacer
13 Rubber bush
14 Special washer
15 Bolt

Fig. 4.6 Rear suspension unit – XR models

Fig. 4.7 Rear suspension unit – XL models

1   Gas-filled damper
2   Spring guide
3   Bump stop
4   Spring
5   Spring adjuster
6   Lower spring seat
7   Nut
8   Lower mounting lug
9   Rubber bush – 2 off
10 Spacer

13.4a Compress spring and displace slotted spring seat

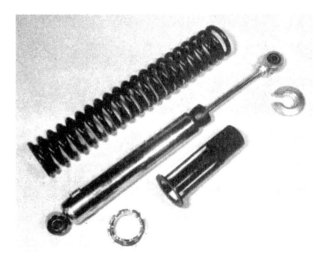

13.4b The rear suspension unit components

14 Swinging arm rear forks: dismantling, examination and renovation

1   The swinging arm fork pivots on a pair of headed bushes which are pressed into the fork crossmember. The internal surfaces of the bushes bear upon a hard steel sleeve, which in turn is secured by a long pivot bolt between the frame bosses.

2   Wear will inevitably take place in the bushes after a period of time, although regular greasing, using the grease nipple provided, will greatly extend their life. If greasing gets put off indefinitely, as often happens, the bearing surfaces will dry up allowing rapid wear to take place.

3   One of the first symptoms of wear is a sensation of vagueness in the machine's handling. Curiously, it is often the front forks which are blamed initially, as the problem seems to be one of imprecise steering. Pronounced wear will allow the swinging arm and rear wheel to jump from one side of the extent of play to the other. Such a machine is very far from roadworthy, and should not be used until an overhaul has taken place.

4   An indication of wear can be found by placing the machine vertically, blocking it securely in a suitable position so that the rear wheel is clear of the ground. The fork can be pushed and pulled to the side and any slight wear in the bushes will be proportionally exaggerated at the fork ends. Any discernible play warrants renewal of the worn parts.

5   Commence by securing the machine so that it is in no danger of toppling over.

6   Remove the rear wheel as described in Chapter 5, Section 6. Detach the upper chainguard by removing the single front bolt and the two bolts at the rear. The lower guard can be removed, if required, to prevent damage.

7   Remove the lower mounting bolt from each of the rear suspension units, which can then be pivoted upwards to clear the swinging arm. Slacken and remove the pivot shaft nut. Support the swinging arm and withdraw the spindle. As the swinging arm is drawn away from the frame the end caps will probably drop away, and they should be retrieved and placed to one side.

8   Displace the hardened steel sleeve, and wash the swinging arm bore, bushes and sleeve in clean petrol (gasoline) or degreasing solution. Check the internal bore of the bushes for wear or scoring, and likewise the steel sleeve for wear and corrosion. Dimensions are given for the various rubbing surfaces in the Specifications Section, and if desired, these parts can be measured. In practice, the steel sleeve can be examined visually, renewal being necessary if obvious wear or damage has taken place. The bushes should be renewed as a matter of course if any trace of movement has been detected. Note that the bushes will certainly be damaged during removal, and therefore cannot be reused once they have been driven out of the swinging arm.

9   The bushes can be removed by passing a long drift or bar through the swinging arm bore, and driving out the bush on the opposite side. Take care to support the swinging arm to avoid any risk of damage. When fitting new bushes, place a flat plate across the head of the bush to spread the impact evenly, and ensure that the bush enters the bore squarely.

10   Reassemble the swinging arm by reversing the dismantling sequence. The steel sleeve should be coated with high melting point grease. Do not omit to refit the end caps to the swinging arm before installing it in the frame. The end caps can be held in position with grease, or by passing a screwdriver into the bore through the frame lug. Tighten the swinging arm pivot shaft nut to the appropriate torque setting.

Swinging arm pivot torque setting

XL250S and XR250	*5.5 – 7.0 kgf m (40 – 51 lbf ft)*
XL500S and XR500	*7.0 – 10.0 kgf m (51 – 72 lbf ft)*

15 Prop stand: examination

1   The prop stand, or side stand, consists of a steel leg attached to the left-hand side of the frame by means of a pivot bolt which passes through a frame lug. An extension spring holds the stand in a horizontal, retracted, position when not in use. Although it is a simple assembly, care should be taken to ensure that it is in good condition, as it provides the machine's sole means of support in the absence of the rider.

2   Check that the pivot bolt is secured and that the extension spring is in good condition and not overstretched. An accident is almost certain if the stand extends whilst the machine is on the move.

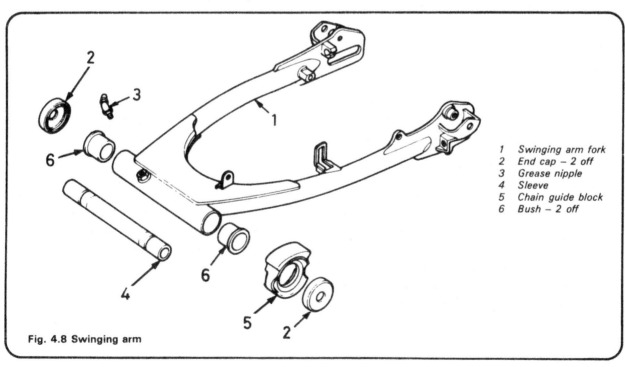

1   Swinging arm fork
2   End cap – 2 off
3   Grease nipple
4   Sleeve
5   Chain guide block
6   Bush – 2 off

Fig. 4.8 Swinging arm

14.6 Remove chainguard to facilitate dismantling

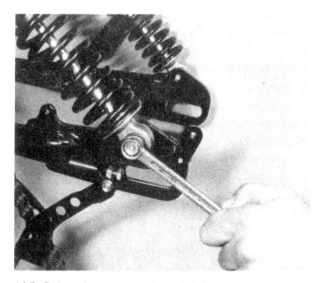

14.7a Release lower suspension unit bolts

14.7b Remove swinging arm pivot bolt...

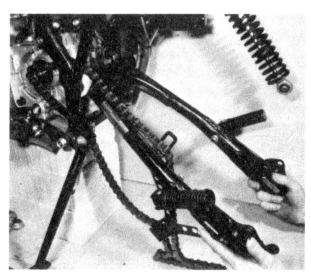

14.7c ...and lift swinging arm clear of frame

14.8a Release end caps from ends of cross tube

14.8b Displace the hardened steel inner sleeve

14.8c The pivot bushes can be drifted out if worn

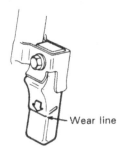

Wear line

Fig. 4.9 Side stand pad wear limit

16 Footrests: examination and renovation

1   The front footrests are bolted to lugs welded to the frame lower engine tubes. The footrests pivot upwards on their mounting brackets and are spring loaded to keep them in their horizontal position. If an obstacle is struck they will fold upwards, reducing the risk of damage to the rider's foot or to the main frame.

2   If the footrests are damaged in an accident, it is possible to dismantle the assembly into its component parts. Detach each footrest from the frame lugs and separate the folding foot piece from the bracket on which it pivots by withdrawing the split pin and pulling out the pivot shaft. It is preferable to renew the damaged parts, but if necessary, they can be bent straight by clamping them in a vice and heating to a dull red with a blow lamp whilst the appropriate pressure is applied. Do not attempt to straighten the footrests while they are attached to the frame.

17 Rear brake pedal: examination and renovation

1   The rear brake pivots on a special bolt which screws into the right-hand footrest plate. If it proves necessary to remove the pedal it will be first be necessary to slacken off the rear brake adjuster so that the cable can be detached, and to release the single bolt which secures the footrest assembly. As the latter is removed, disconnect the new brake switch operating spring.

2   The pedal can be removed from the footrest bracket by unscrewing its pivot bolt. This passes into the inner face of the

14.10 Do not omit chain guide block during reassembly

bracket and also supports the pedal return spring, which should be disengaged. If removed, the pivot bolt should always be released so that the pedal can be inspected and lubricated. Clean out any old grease, then re-grease the pedal bore. When reassembling the pedal, note the dust seal and cover which is fitted on each side of the pedal bore.

3   Installation is a reversal of the removal sequence, remembering to refit the brake switch spring as the assembly is offered up. The appropriate torque settings are given below, where these are available.

Rear brake pivot bolt– torque setting

XL250S	*Not available*
XR250	*Not available*
XL500S	*5.5 – 8.0 kgf m (40 – 58 lbf ft)*
XR500	*5.5 – 7.0 kgf m (40 – 58 lbf ft)*

Footrest mounting bolt – torque setting

XL250S	*Not available*
XR250	*Not available*
XL500S	*7.0 – 10.0 kgf m (51 – 72 lbf ft)*
XR500	*7.0 – 10.0 kgf m (51 – 72 lbf ft)*

18 Speedometer head: removal and examination

1   The speedometer is a self-contained unit on XR models, whilst on the XL versions it forms part of the instrument panel assembly. The removal sequence for each type is described in Section 4, paragraphs 3 and 9.

2   Apart from defects in either the drive or drive cables, a speedometer which malfunctions is difficult to repair. Fit a replacement or alternatively entrust the repair to a competent instrument repair specialist.

3   Remember that a speedometer in correct working order is a statutory requirement in the UK. Apart from this legal necessity, reference to the odometer readings is the most satisfactory means of keeping pace with the maintenance schedules.

19 Speedometer drive cable: examination and maintenance

1   If the operation of the speedometer becomes jerky or sluggish, the drive cable should be the first area checked for wear or damage. Check that the cable has not become trapped or kinked internally. This is best done by removing the cable completely and then turning the inner cable. If a tight spot is found, it is probably the cause of erratic operation of the instrument.

2   If the speedometer stops working suspect a broken drive cable unless the odometer readings continue. Inspection will show whether the inner cable has broken, in which case renewal of the complete cable will be necessary. The inner cable cannot be obtained separately.

20 Speedometer drive: location and examination

1   The speedometer cable is driven by a worm mechanism incorporated in the front of the brake plate. The assembly is of robust construction and requires no regular maintenance. Apart from obvious breakages, which will require renewal of the parts concerned it will suffice to grease the drive mechanism whenever attention is given to the front wheel bearings.

21 Cleaning the machine

1   After removing all surface dirt with a rag or sponge which is washed frequently in clean water, the machine should be allowed to dry thoroughly. Application of car polish or wax to the cycle parts will give a good finish, particularly if the machine received this attention at regular intervals.

2   The plated parts should require only a wipe with a damp rag, but if they are badly corroded, as may occur during the winter when the roads are salted, it is permissible to use one of the proprietary chrome cleaners. These often have an oily base which will help to prevent corrosion from recurring.

3   If the engine parts are particularly oily, use a cleaning compound such as Gunk or Jizer. Apply the compound whilst the parts are dry and work it in with a brush so that it has an opportunity to penetrate and soak into the film of oil and grease. Finish off by washing down liberally, taking care that water does not enter the carburettor, air cleaner or the electrics.

4   If possible, the machine should be wiped down immediately after it has been used in the wet, so that it is not garaged in damp conditions which will promote rusting. Make sure that the chain is wiped and re-oiled, to prevent water from entering the rollers and causing harshness with an accompanying rapid rate wear. Remember there is less chance of water entering the control cables and causing stiffness if they are lubricated regularly as described in the Routine Maintenance Section.

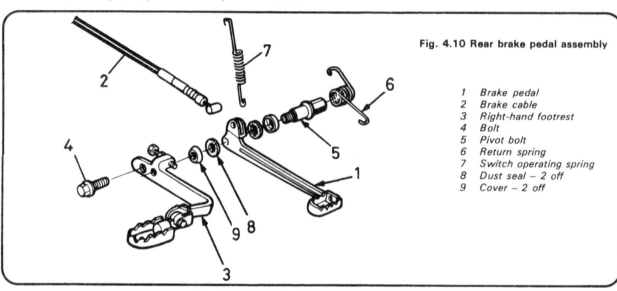

Fig. 4.10 Rear brake pedal assembly

1   Brake pedal
2   Brake cable
3   Right-hand footrest
4   Bolt
5   Pivot bolt
6   Return spring
7   Switch operating spring
8   Dust seal – 2 off
9   Cover – 2 off

22 Fault diagnosis: frame and forks

Symptom	Cause	Remedy
Machine veers either to the left or the right with hands off handlebars	Bent frame Twisted forks Wheels out of alignment	Check and renew. Check and renew. Check and re-align.
Machine rolls at low speed	Overtight steering head bearings	Slacken until adjustment is correct.
Machine judders when front brake is applied	Slack steering head bearings Worn fork legs	Tighten until adjustment is correct. Dismantle forks and renew worn parts.
Machine pitches on uneven surfaces	Ineffective fork dampers Ineffective rear suspension units Suspension too soft	Check oil content. Check whether units still have damping action. Raise suspension unit adjustment one notch.
Fork action stiff	Fork legs out of alignment (twisted in yokes)	Slacken yoke clamps, and fork top bolts. Pump for several times then retighten from bottom upwards.
Machine wanders. Steering imprecise. Rear wheel tends to hop	Worn swinging arm pivot	Dismantle and renew bushes and pivot shaft.

Chapter 5 Wheels, brakes and tyres

For modifications and information relating to later models, see Chapter 7

Contents

Specifications

	250 models	500 models
Tyres		
Size:		
Front	3.00X23	3.00X23
Rear	4.60X18	4.60X18

Note: XL models are fitted with 4-ply (4PR) tyres, and XR models 6-ply (6PR) tyres

Pressures:*		
Front	21 psi (1.5 kg cm^2)	14 psi (1.0 kg cm^2)
Rear:		
Solo	21 psi (1.5 kg cm^2)	17 psi (1.0 kg cm^2)
With passenger	25 psi (1.75 kg cm^2)	—

Note: pressures shown are for road use (XL models) or off-road use (XR models). Pressures may be varied for off-road work, but should be restored to normal when machines are ridden on the public road.

Brakes	XL250 S	XR250	500 models
Type	Internal expanding drum brakes front and rear (all models)		
Swept area:			
Front	109 cm^2	64 cm^2	110 cm^2
	(17.0 sq in)	(9.9 sq in)	(17.1 sq in)
Rear	103 cm^2	60 cm^2	122.5 cm^2
	(16.0 sq in)	(9.3 sq in)	(19.0 sq in)

1 General description

The wheels used on all the models covered by this manual are of similar construction. All are of the wire-spoked type and incorporate single leading shoe (sls) drum brakes front and rear. Unlike most manufacturers, Honda have chosen to use a 23 inch front wheel in preference to the more conventional 21 inch type. Although this is claimed to give better clearance and control, it does make the choice of replacement tyres somewhat limited. The original tyre fitments are variations of the Honda 'claw action' types, those of the XL models being designed to provide good adhesion on the road as well as off-road.

2 Front wheel: examination and renovation

1   Place the machine on a suitable support so that the front wheel is raised clear of the ground. Spin the wheel and check the rim alignment. Small irregularities can be corrected by tightening the spokes in the affected area although a certain amount of experience is necessary to prevent over-correction. Any flats in the wheel rim will be evident at the same time. These are more difficult to remove and in most cases it will be necessary to have the wheel rebuilt on a new rim. Apart from the effect on stability, a flat will expose the tyre bead and walls to greater risk of damage if the machine is run with a deformed wheel.

2   Check for loose and broken spokes. Tapping the spokes is the best guide to tension. A loose spoke will produce a quite different sound and should be tightened by turning the nipple in an anticlockwise direction. Always check for run out by spinning the wheel again. If the spokes have to be tightened by an excessive amount, it is advisable to remove the tyre and tube as detailed in Section 12 of this Chapter. This will enable the protruding ends of the spokes to be ground off, thus preventing them from chafing the inner tube and causing punctures.

3   On any machine used for trail or enduro work, particular emphasis must be placed on the importance of regular cleaning and inspection of the wheels, paying particular attention to the wheel rim as the likelihood of buckles or dents is much greater with a machine used off-road. The need for retightening spokes will also be greater on these machines, and remedial action must be taken to avoid rapid deterioration or failure of the wheel.

3   Front wheel: removal and refitting

1   As no centre stand is fitted, some alternative means of supporting the machine with the wheel clear of the ground must be found prior to wheel removal. A stout wooden box or similar is usually suitable, but ensure that there is no risk of the machine toppling whilst it is being worked on. Commence dismantling by slackening off the front brake cable adjustment at both ends, then disconnect the cable at the wheel after disengaging the cable at the operating lever. The speedometer drive cable should be released by its single retaining screw.

2   Straighten and remove the split pin (XL models only) which secures the wheel spindle nut, then remove the nut itself. Remove the clamp bolt(s) at the other end of the spindle, then withdraw the spindle and lower the wheel clear of the forks.

3   With the wheel removed from the machine, the brake backplate assembly can be removed for examination. Refer to next section for brake inspection.

3   The wheel may be refitted by reversing the removal procedure. When lifting the wheel into position ensure that the lug on the left-hand fork leg engages with the slot in the brake backplate. If this requirement is overlooked the brake will lock on at the first application. Insert the wheel spindle and tighten the castellated nut before tightening the single clamp bolt (XL models) or four clamp nuts (XR models). In the latter case check that the spindle clamp is fitted with the 'up' mark in the appropriate position. Reconnect the speedometer cable and the brake cable and then adjust the brake. Finally, check that a new split pin has been fitted to secure the spindle nut (where fitted).

3.1a Disconnect brake cable from brake plate and fork leg (arrowed)

3.1b Speedometer cable is secured by a single screw

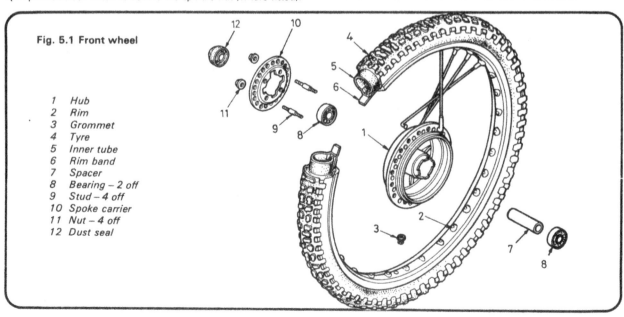

Fig. 5.1 Front wheel

1   Hub
2   Rim
3   Grommet
4   Tyre
5   Inner tube
6   Rim band
7   Spacer
8   Bearing – 2 off
9   Stud – 4 off
10 Spoke carrier
11 Nut – 4 off
12 Dust seal

3.2a Straighten and remove split pin, then release nut

3.2b Remove the wheel spindle pinch bolt...

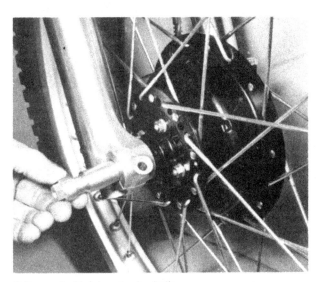

3.2c ...and withdraw wheel spindle

4 Front drum brake: removal, examination and renovation

1   Before attention is turned to the brake components, removal of the front wheel will be required. This may be
accomplished by following the sequence in the preceding
Section.
2   With the wheel removed from the machine the brake
backplate assembly can be detached from the wheel. The
backplate together with the shoes will lift out of the drum quite
easily. Any dust in the drum should be wiped out immediately
using a damp rag. Note that the dust contains a proportion of
asbestos. This presents a health hazard if inhaled and **must not**
be blown out with compressed air.
4   Examine the condition of the brake linings. If they are thin
or unevenly worn, the brake shoes should be renewed. The
linings are bonded on and cannot be supplied separately. The
linings are 4 mm (0.16 inch) thick when new and should receive
attention when reduced to the wear limit thickness of 2 mm
(0.08 inch).
5   To remove the brake shoes, turn the brake operating lever
so that the brake is in the fully on position. Pull the brake shoes

apart against their return spring pressure to free them from their
operating cams, after withdrawing the split pins used to anchor
their ends. The shoes can then be lifted away, complete with
the return springs by reverting to a 'V' formation. When they are
clear of the brake backplate, the return springs can be removed
and the shoes separated.
6   Before replacing the brake shoes, check that the brake
operating cam is working smoothly and is not binding in the
pivot bush. To remove the cam for greasing, detach the
operating arm from the splined end of the shaft, after marking
the shaft and the arm so that they are replaced in an identical
position. The shaft will then push through inwards. Note that
the brake arm return spring and wear indicator pointer, will be
displaced as the cam is removed.
7   Check the inner surface of the brake drum. The surface on
which the brake shoes operate should be smooth and free from
score marks or indentations, otherwise reduced braking efficiency will be inevitable. Remove all traces of brake lining dust and
wipe with a clean rag soaked in petrol to remove all traces of
grease and oil. The brake drum does not normally wear
noticeably, but if scoring necessitates having the drum surface
skimmed on a lathe, the finished diameter must not exceed
141.0 mm (5.55 in). The nominal diameter of the drum is 14.0
mm (5.15 in).
8   To reassemble the brake shoes on the backplate, fit the
return springs and pull the shoes apart, holding them in a 'V'
formation. If they are now located with the brake operating
cams and pivots, they can be pushed back into position by
pressing downwards in order to snap them into position. Do not
use excessive force, or there is risk or distorting the brake shoes
permanently.
9   After the brake has been reassembled and the wheel
installed, it is important that the cable slack is adjusted. Set the
handlebar lever adjuster a few turns from the fully closed
position. The adjuster at the lower end of the cable should now
be arranged so that the wheel is just free to turn. Brake
operation should commence just after the handlebar lever starts
to move, thus minimising any delay in operation. Tighten the
locknut and refit the rubber boot. The handlebar lever adjuster
can now be used to compensate for the gradual wear which will
take place in the lining material.
10 If new brake shoes have been fitted, they should be allowed
to bed in for the first few hundred miles. Note that during this
time braking efficiency will be impaired until the new shoes
conform exactly to the drum surface. Avoid hard application
during the bedding in period to prevent the lining surface
becoming glazed.

4.2 Brake plate assembly can be lifted out of drum

4.4 Measure thickness of brake lining material

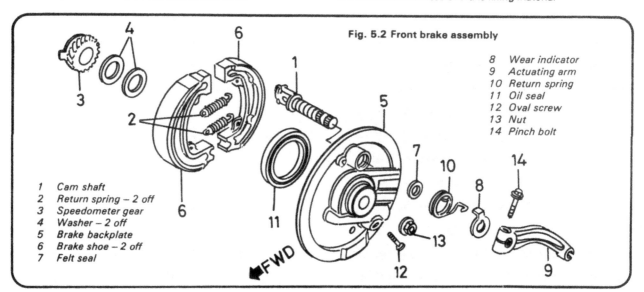

Fig. 5.2 Front brake assembly

1   Cam shaft
2   Return spring – 2 off
3   Speedometer gear
4   Washer – 2 off
5   Brake backplate
6   Brake shoe – 2 off
7   Felt seal

8   Wear indicator
9   Actuating arm
10 Return spring
11 Oil seal
12 Oval screw
13 Nut
14 Pinch bolt

FWD

5   Front wheel bearings: examination and replacement

1   The front wheel is supported on two journal ball bearings with sealed outer faces. These are packed with grease during assembly, and under normal conditions can be expected to last a considerable time. Bearing failure is normally due to the ingress of water or dust which causes corrosion or damage to the bearing surfaces, and this is unlikely to occur unless the seal has failed.

2   Access to the bearings is gained after the front wheel has been removed and the brake backplate assembly lifted away, as described in the preceding sections. Place the wheel on the workbench, raising the hub clear of the surface by placing wooden blocks on each side of the bearing boss. The bearing nearest the workbench can now be driven out by passing a drift down through the hub. The tubular spacer which is fitted between the bearings can be displaced slightly to allow the drift to engage on the lower race. When the lower bearing has been knocked free, remove the spacer and invert the wheel to allow the remaining bearing to be driven out.

3   Remove the old grease from the bearings by washing them in petrol. When clean and dry, check the condition of each bearing by rotating it to check for roughness and by feeling for

play. If there is any doubt about their condition, renew them.

4   When fitting new or sound bearings, pack them and the hub with high melting point grease prior to assembly. The bearings should be fitted with their sealed faces outwards, using a large socket or similar to drive them squarely into the hub.

6   Rear wheel: examination, removal and replacement

1   Place the machine on a stand or blocks so that the rear wheel is raised clear of the ground. Check the tension of the spokes and trueness of the wheel rim in the same way as described for the front wheel in Section 2. The remarks and recommendations given in this Section can be applied to the rear wheel, noting that care must be taken to avoid confusing wheel bearing play with wear that may have developed in the swinging arm pivot.

2   To remove the rear wheel, start by releasing the rear brake cable by unscrewing the adjuster at the cable end. Free the cable and spring and displace the trunnion in the end of the brake arm. Fit these components to the end of the cable and retain them with the adjuster nut to prevent loss.

3   On those machines fitted with a spring link in the final drive chain the chain may be separated after removal of the link and

run off the wheel sprocket. On other machines slacken the chain adjuster locknuts and unscrew the adjuster bolts. This will give enough slack in the chain, after wheel spindle removal, to allow the chain to be disengaged from the sprocket.

4   Straighten and remove the split pin from the wheel spindle end, remove the nut and withdraw the spindle. On 500 models detach the brake torque arm locating plate from the boss on the brake backplate and detach the end of the torque arm. Where necessary move the wheel forwards and disengage the chain from the sprocket. The wheel can now be moved to the rear away from the fork ends.

5   Refit the wheel by reversing the dismantling procedure. On 250 models ensure that the slot in the brake backplate locates with the lug on the swinging arm right-hand fork tube. On 500 models locate the torque arm end on the backplate boss, retaining it in position by means of the retaining plate. On those machines fitted with a rear chain spring link the chain adjusters will not have been disturbed but the rear wheel alignment should be checked. On other models, chain adjustment and wheel alignment should be carried out. Refer to Section 12 for details. Ensure that a new split pin is fitted to the wheel spindle after the nut is tightened fully. Adjust the rear brake. Ensure that when the chain spring link is fitted it is positioned so that the closed end of the horse-shoe spring is pointing in the direction of normal chain travel.

5.4a Bearings are fitted with sealed face outwards

5.4b A large socket can be used to drive bearings home

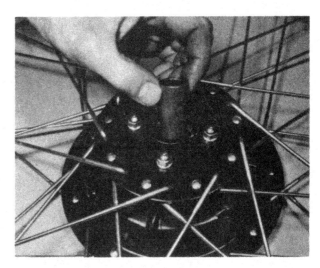

5.4c Do not omit the tubular spacer between bearings

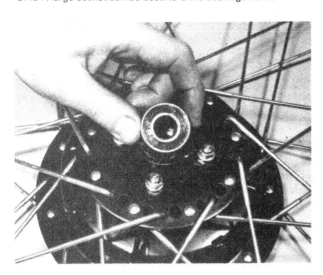

5.4d Fit the remaining bearing...

5.4e ...followed by seal

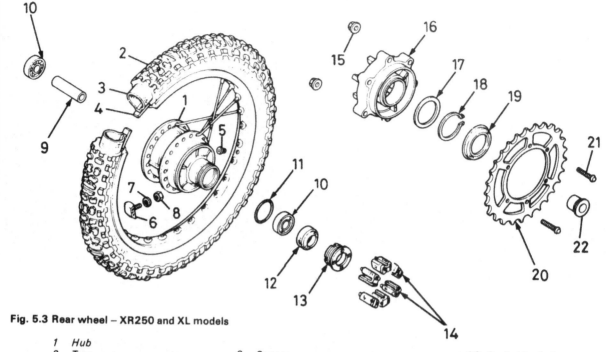

Fig. 5.3 Rear wheel – XR250 and XL models

1	Hub		
2	Tyre	9 Spacer	16 Cush drive hub
3	Inner tube	10 Bearing – 2 off	17 Thrust washer
4	Rim band	11 O-ring	18 Circlip
5	Grommet	12 Dust seal	19 Dust seal
6	Security bolt	13 Bearing retainer	20 Sprocket
7	Spring washer	14 Cush drive rubber – 6 off	21 Bolt – 6 off
8	Nut	15 Nut – 6 off	22 Collar

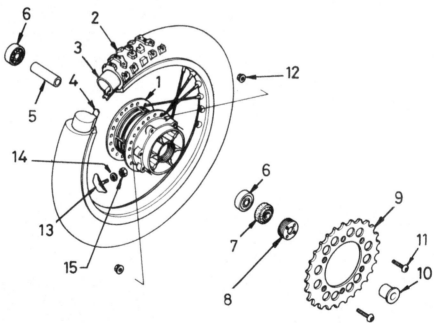

Fig. 5.4 Rear wheel – XR 500 model

1	Hub	6 Bearing – 2 off	11 Bolt – 6 off
2	Tyre	7 Dust seal	12 Nut – 6 off
3	Inner tube	8 Bearing retainer	13 Security bolt
4	Rim band	9 Sprocket	14 Spring washer
5	Spacer	10 Collar	15 Nut

6.4a Straighten the split pin and withdraw it

6.4b Remove rear wheel spindle nut

6.4c Withdraw the spindle to release wheel

6.4d Special adjuster sleeve will be displaced

7   Rear brake: examination and renovation

1   The rear brake assembly can be removed after the rear wheel has been released from the machine as described in the preceding Section. The rear brake is, for all practical purposes, identical to the front unit, and the same examination and overhaul procedures can be applied. The brake lining thickness and drum diameter is the same as that of the front brake. See Section 3 for details.

8   Cush drive assembly: examination and renovation – except XR500

1   Whilst the rear wheel is removed from the frame, it is convenient to examine the cush drive assembly. It comprises six vanes that form an integral part of a plate bolted to the rear of the final drive sprocket, which makes contact with six specially shaped rubber pads that are a push fit over projections cast on the inside of the rear wheel hub. The drive is transmitted via these rubbers, which permit the sprocket to move within certain limits and cushion any surges or roughness in the transmission which would otherwise create the impression of harshness.

2   After a lengthy period of service, the rubbers will com-

mence to break up or otherwise deteriorate. If any signs of damage are noted, the rubbers should be renewed, as a complete set. It is unlikely that the vanes will require attention since they are not normally subject to wear. Excessive movement of the rear wheel sprocket is usually a sign that the cush drive assembly is in need of attention.

3   To remove the cush drive and sprocket unit, prise off the dust seal which covers the wheel hub extension to reveal the large circlip and thrust washer which retain the cush drive unit. The unit can be pulled off after the circlip and washer have been removed.

9   Rear wheel bearings: examination and overhaul

1   If there has been evidence of play in the rear wheel bearings it will be necessary to remove them for further examination and possible renewal. Start by removing the rear wheel from the machine (Section 6) and, where fitted, the cush drive and sprocket assembly from the wheel (Section 8). It will be necessary to use some form of special tool to remove the bearing retainer from the sprocket side of the wheel. In the absence of the appropriate Honda tool, this can be improvised as shown in the accompanying photographs. Note that the retainer will have been staked into the hub during assembly and

may prove very hard to remove. If this situation arises it is safer to entrust the work to a Honda dealer than to risk damage to the retainer or hub threads.

2   When the retainer has been removed, the wheel bearings can be driven out of the hub using the procedure given for the front wheel. Remove all the old grease from the hub and bearings, giving the latter a final wash in petrol. Check the bearings for play or any signs of roughness as they are rotated. If there is any doubt about their condition, renew them.

3   Before driving the bearings back into the hub, pack the hub with new grease and also grease the bearings. Use a tubular drift against the outer races to drive them back into position, not forgetting the hollow distance piece between them.

4   Fit and tighten the threaded retainer. Once in position the retainer should be staked in place using a sharp centre punch. This will obviate any risk of it slackening in use. A thread locking compound may be used in preference to staking. Complete assembly by reversing the removal sequence. When the rear wheel has been fitted back into the frame, check brake adjustment and chain tension before use.

7.1a Lift brake plate assembly clear of drum

7.1b Brake shoe arrangement is similar to front unit

7.1c Check drum surface for wear and scoring

8.2 Damper rubbers will allow excessive play if worn

8.3a Remove dust seal from hub boss

8.3b Remove circlip...

8.3c ...and plain washer...

8.3d ...to allow cush drive hub to be released

9.1a Withdraw spacer from hub assembly

9.1b Improvised tool to facilitate retainer removal

9.4 Retainer should be staked in position

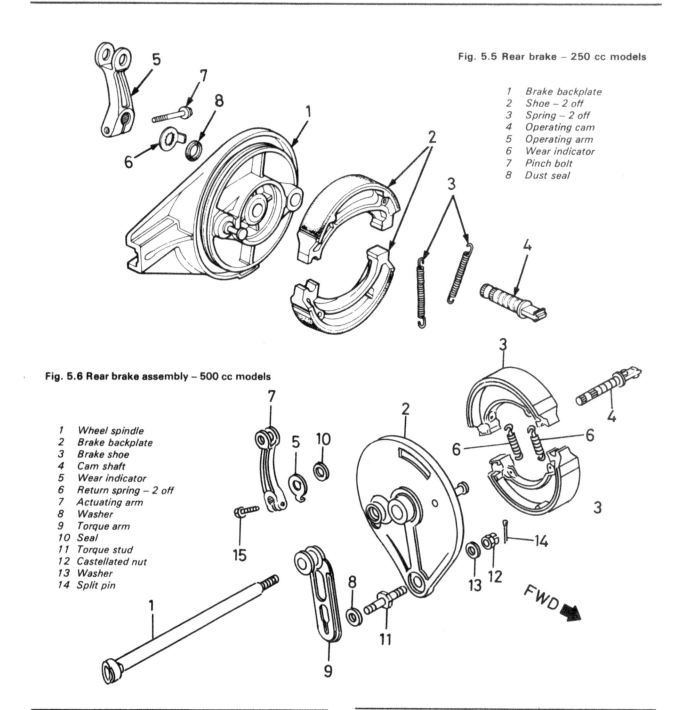

Fig. 5.5 Rear brake – 250 cc models

1   Brake backplate
2   Shoe – 2 off
3   Spring – 2 off
4   Operating cam
5   Operating arm
6   Wear indicator
7   Pinch bolt
8   Dust seal

Fig. 5.6 Rear brake assembly – 500 cc models

1   Wheel spindle
2   Brake backplate
3   Brake shoe
4   Cam shaft
5   Wear indicator
6   Return spring – 2 off
7   Actuating arm
8   Washer
9   Torque arm
10 Seal
11 Torque stud
12 Castellated nut
13 Washer
14 Split pin

10 Rear brake adjustment

1   The rear brake adjustment should be checked whenever the rear wheel has been removed or disturbed, or to take up normal wear. Start by setting the brake pedal height to the most convenient position by means of the stop bolt and locknut at the pedal.

2   The brake adjustment is set by turning the adjuster nut at the wheel end of the rear brake cable. The position is largely a matter of personal choice and will usually be arranged so that brake operation commences about 10 – 15 mm ($\frac{3}{8}$ – $\frac{5}{8}$ in) from the rest position. The play is measured at the pedal end. Check that the brake light switch is set to come on as the pedal is depressed. This is covered in Chapter 6.

11 Rear wheel sprocket: removal, examination and replacement

1   The rear wheel sprocket is secured by six screws and nuts either directly to the wheel hub (XR500), or to the cush drive hub (all other models). Prior to removing the sprocket the wheel must be detached from the frame as detailed in Section 5, and where necessary, the cush drive hub removed after displacing the dust seal and large circlip.

2   The sprocket should be renewed if the teeth are hooked, chipped, broken or badly worn. It is considered bad practice to renew one sprocket on its own; both drive sprockets should be renewed as a pair, preferably with a new final drive chain. If this recommendation is not observed, rapid wear resulting from the

running of old and new parts together will necessitate even earlier replacement on the next occasion.

3   Owners of XL machines may find that they wish to alter the overall gearing to suit the prevailing operating conditions. The standard gearing is designed as a reasonable compromise between road and off-road gearing, but specific owners may use the machine predominantly for one or the other application, and consequently wish to change the gearing accordingly. No alternative sprockets are produced by the manufacturers, but careful perusal of the motorcycle newspapers and magazines will unearth the addresses of various specialist manufacturers, some of whom will undertake to produce one-off sprockets if they are not normally stocked. A similar approach may be applied to XR models where a change of gearing would help cope with particular types of terrain.

12 Final drive chain: examination and lubrication

1   The final drive chain is exposed for most of its travel and has only a lightweight chainguard to protect the upper run. No provision is made for lubricating the chain.
2   The chain tension will require adjustment at regular intervals, especially if the machine is used for competition or off-road riding. This is accomplished by slackening the rear wheel after first removing the split pin through the end of the spindle, and then moving the rear wheel backwards by unscrewing the two chain adjusting bolts that bear on the wheel spindle. The locknut of each adjuster should be slackened first, and re-tightened after adjustment has been effected. Turn each adjuster an equal amount, using the scribe marks on the fork ends to provide a visual check.
3   Chain tension on the XL models is correct if there is 30 – 40 mm ($1\frac{1}{4}$ – $1\frac{5}{8}$) free play at the centre of the lower run, whilst on XR versions there must be at least 20 mm ($\frac{3}{4}$ in) between the lower run of the chain at the tensioner and the underside of the swinging arm. The normal clearance at this point should be 30 – 40 mm ($1\frac{1}{4}$ – $1\frac{5}{8}$ in) approximately. Do not run the chain overtight to compensate for uneven wear. A tight chain will place excessive stresses on the gearbox and rear wheel bearings, leading to their early failure. It will also absorb a surprising amount of power.
4   After a period of running, the chain will require lubrication. Lack of oil will accelerate the rate of wear of both chain and sprockets and will lead to harsh transmission. Lubricate the chain whilst it is in place on the machine using one of the proprietary aerosol chain greases. This is recommended rather than engine oil because it has a higher viscosity and is less likely to be flung off the fast-moving chain. XR models are fitted with a chain with a spring link which allows removal of the chain for a thorough clean in a petrol/paraffin mix. The chain can then be immersed in one of the special chain greases such as 'Linklyfe' or 'Chainguard'. XL models are fitted with a continuous chain, the spring link being omitted in the interests of greater chain strength. The chain can only be removed after removal of the swing arm fork which requires a large amount of dismantling work. When the time comes for chain renewal some thought might be given to fitting a replacement chain with a spring link; this will allow easy removal for thorough cleaning and lubrication which will prolong the life of the chain.
5   Remember that if the machine is used for competition work or in particularly dusty conditions, the chain will require much more frequent attention if it is not to wear rapidly.
6   To check whether the chain is due for replacement, lay it lengthwise in a straight line and compress it endwise until all play is taken up. Anchor one end, then pull in the opposite direction to take up the play which develops. If the chain extends by more than $\frac{1}{4}$ inch per foot, it should be replaced in conjunction with the sprockets. Note that this check should ALWAYS be made after the chain has been washed out, but before any lubricant is applied, otherwise the lubricant may take up some of the play. It is evident that this approach to chain wear inspection can only, practically, be applied to chains with

a spring link. Some idea of the condition of a continuous chain can be gauged by applying tension to the chain by pulling up on the lower run. If, with the chain so tensioned, the chain can be pulled away from the sprocket more than about half a link, then there is evidence that chain renewal is required. Some experience is required when determining chain wear by this method.
7   If desired, wheel alignment can be checked by running a plank of wood parallel to the machine so that it is equidistant from either side of the front wheel tyre when tested on both sides of the rear wheel. It will not touch the front tyre because this tyre has a smaller cross section. See the accompanying diagram.
8   The chain fitted as standard is of Japanese manufacture. When renewal becomes necessary, it should be noted that a Renold equivalent, of British manufacture, is available. When purchasing a replacement, take along the old chain as a pattern or, if known, a note of the size and number of pitches.

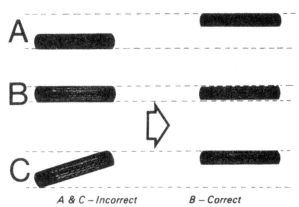

A & C – Incorrect   B – Correct

Fig. 5.7 Wheel alignment

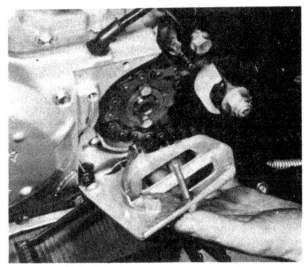

12.6 Check sprocket condition by releasing cover

13 Tyres – removal and replacement

1   At some time or other the need to arise to remove and replace the tyres, either as the result of a puncture or because a replacement is required to offset wear. To the inexperienced, tyre changing represents a formidable task yet if a few simple rules are observed and the technique learned, the whole operation is surprisingly simple.

2   To remove the tyre from either wheel, first detach the wheel from the machine by following the procedure in the relevant section of this Chapter, depending on whether the front or rear wheel is involved. Deflate the tyre by removing the valve insert and when it is fully deflated, push the bead of the tyre away from the wheel rim on both sides so that the bead enters the centre well of the rim. Remove the locking ring and push the tyre valve into the tyre itself. All models have security bolts, which are special bolts with extended moulded flanges arranged to clamp the tyre bead to the rim. On a road machine of this capacity, security bolts are not normally fitted, as there is no real need for them, but on an off-road machine, the acceleration and braking forces are such that the tyre tends to 'creep' on the rim. If the tyre is left unanchored, it would eventually tear the valve from the inner tube.

3   The front tyre has one bolt, whilst the rear tyre, which transmits the engine power, is fitted with two. It will be necessary to slacken the clamping nut(s) and push the bolt(s) up into the tyre before the tyre can be removed. Insert a tyre lever close to the valve and lever the edge of the tyre over the outside of the wheel rim. Very little force should be necessary; if resistance is encountered it is probably due to the fact that the tyre beads have not entered the well of the wheel rim all the way round the tyre.

4   Once the tyre has been edged over the wheel rim, it is easy to work around the wheel rim so that the tyre is completely free on one side. At this stage, the inner tube can be removed.

5   Working from the other side of the wheel, ease the other edge of the tyre over the outside of the wheel rim that is furthest away. Continue to work around the rim until the tyre is free completely from the rim.

6   If a puncture has necessitated the removal of the tyre, reinflate the inner tube and immerse it in a bowl of water to trace the source of the leak. Mark its position and deflate the tube. Dry the tube and clean the area around the puncture with a petrol soaked rag. When the surface has dried, apply the rubber solution and allow this to dry before removing the backing from the patch and applying the patch to the surface.

7   It is best to use a patch of self-vulcanising type which will form a very permanent repair. Note that it may be necessary to remove a protective covering from the top surface of the patch, after it has sealed in position. Inner tubes made from synthetic rubber may require a special type of patch and adhesive if a satisfactory bond is to be achieved.

8   Before replacing the tyre, check the inside to make sure the agent which caused the puncture is not trapped. Check also the outside of the tyre, particularly the tread area, to make sure nothing is trapped that may cause a further puncture.

9   If the inner tube has been patched on a number of past occasions, or if there is a tear or large hole, it is preferable to discard it and fit a replacement. Sudden deflation may cause an accident, particularly if it occurs with the rear wheel.

10 To replace the tyre, inflate the inner tube sufficiently for it to assume a circular shape but only just. Then push it into the tyre so that it is enclosed completely. Lay the tyre on the wheel at an angle and insert the valve captive in its correct location.

11 Starting at the point furthest from the valve, push the tyre bead over the edge of the wheel rim until it is located in the central well. Continue to work around the tyre in this fashion until the whole of one side of the tyre is on the rim. It may be necessary to use a tyre lever during the final stages.

12 Make sure there is no pull on the tyre valve and again commencing with the area furthest from the valve, ease the other bead of the tyre over the edge of the rim. Finish with the area close to the valve, pushing the valve up into the tyre until the locking cap touches the rim. This will ensure the inner tube is not trapped when the last section of the bead is edged over the rim with a tyre lever. Ensure that the security bolt(s), where fitted, are pushed up into the tyre to allow the tyre beads to fit below the bolt's flanges. It is important that the inner tube does not become trapped by the security bolt flange, and to this end, the tube should be inflated very slightly.

13 Check that the inner tube is not trapped at any point.

Reinflate the inner tube, and check that the tyre is seating correctly around the wheel rim. There should be a thin rib moulded around the wall of the tyre on both sides which should be equidistant from the wheel rim at all points. If the tyre is unevenly located on the rim, try bouncing the wheel when the tyre is at the recommended pressure. It is probable that one of the beads has not pulled clear of the centre well.

14 Always run the tyres at the recommended pressures and never under or over-inflate. The correct pressure for solo use are given in the Specifications Section of this Chapter. If a pillion passenger is carried, increase the rear tyre pressure only by approximately 4 psi.

15 Tyre replacement is aided by dusting the side walls, particularly in the vicinity of the beads, with a liberal coating of French chalk. Washing up liquid can also be used to good effect, but this has the disadvantage of causing the inner surfaces of the wheel rim to rust.

16 Never replace the inner tube and tyre without the rim tape in position. If this precaution is overlooked there is a good chance of the ends of the spoke nipples chafing the inner tube and causing a crop of punctures.

17 Never fit a tyre which has a damaged tread or side walls. Apart from the legal aspects there is a very great risk of a blowout, which can have serious consequences on any two-wheel vehicle.

18 Tyre valves rarely give trouble, but it is always advisable to check whether the valve itself is leaking before removing the tyre. Do not forget to fit the dust cap which forms an effective second seal.

14 Tyre valve dust caps

1   Tyre valve dust caps are often left off when a tyre has been replaced, despite the fact that they serve an important two-fold function. Firstly, they prevent dirt or other foreign matter from entering the valve and causing the valve to stick open when the tyre pump is next applied. Secondly, they form an effective second seal so that in the event of the tyre valve sticking, air will not be lost. It is particularly important that off road and dual purpose machines are fitted with dust caps to exclude the mud which would otherwise enter and clog the valve.

15 Wheel balance

1   On any high performance machine it is important that the front wheel is balanced, to offset the weight of the tyre valve. If this precaution is not observed, the out-of-balance wheel will produce an unpleasant hammering that is felt through the handlebars at speeds from approximately 50 mph upwards. Whilst this may not cause problems off road, it can become apparent with the XL versions, when they are cruised at fairly high road speeds.

2   To balance the front wheel, place the machine on the centre stand so that the front wheel is well clear of the ground and check that it will revolve quite freely, without the brake shoes rubbing. In the unbalanced state, it will be found that the wheel always comes to rest in the same position, with the tyre valve in the six o'clock position. Add balance weights to the spokes diametrically opposite the tyre valve until the tyre valve is counterbalanced, then spin the wheel to check that it will come to rest in a random position on each occasion. Add or subtract weight until perfect balance is achieved.

3   Only the front wheel requires attention. There is little point in balancing the rear wheel (unless both wheels are completely interchangeable) because it will have little noticeable effect on road holding and general handling.

4   When a new tyre is fitted, the wheel should be rebalanced as a matter of course. Most tyre sidewalls are marked with a spot of coloured paint, denoting the *lightest* point of the tyre. This dot should be positioned immediately adjacent to the tyre valve to counter the out-of-balance forces of the valve. The normal balancing procedure should then be followed.

Tyre changing sequence - tubed tyres

 A Deflate tyre. After pushing tyre beads away from rim flanges push tyre bead into well of rim at point opposite valve. Insert tyre lever adjacent to valve and work bead over edge of rim.

Use two levers to work bead over edge of rim. Note use of rim protectors **B**

 C Remove inner tube from tyre

When first bead is clear, remove tyre as shown **D**

 E When fitting, partially inflate inner tube and insert in tyre

Work first bead over rim and feed valve through hole in rim. Partially screw on retaining nut to hold valve in place. **F**

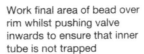 **G** Check that inner tube is positioned correctly and work second bead over rim using tyre levers. Start at a point opposite valve.

Work final area of bead over rim whilst pushing valve inwards to ensure that inner tube is not trapped **H**

16 Fault diagnosis: wheels, brakes and tyres

Symptom	Cause	Remedy
Ineffective brakes	Worn brake lining	Renew.
	Foreign bodies on brake linings surface	Clean.
	Incorrect engagement of brake arm serration	Reset correctly.
	Worn brake cam	Renew.
Handlebars oscillate at low speeds	Buckle or flat in wheel rim, most likely front wheel	Check rim alignment by spinning wheel Correct by retensioning spokes or building on new rim.
	Tyre not straight on rim	Check tyre alignment.
Machine lacks power and poor acceleration	Brakes binding	Warm brake drum provides best evidence. Re-adjust brakes.
Brakes grab when applied gently	Ends of brake shoes not chamfered	Chamfer with file.
	Elliptical brake drum	Lightly skim on lathe
Brake pull-off spongy	Brake cam binding in housing	Free and grease.
	Weak brake shoe springs	Renew if springs have not become displaced.
Harsh transmission	Worn or badly adjusted final drive chain	Adjust or renew
	Hooked or badly worn sprockets	Renew as a pair.
	Loose rear sprocket	Check bolts.
	Worn damper rubbers	Renew rubber inserts.

Chapter 6 Electrical system

For modifications and information relating to later models, see Chapter 7

Contents

Specifications

	XL250S	XL500S	XR models
Generator			
Type	(ac generator) all models		
Charge commences	1500 rpm	1200 rpm	–
Output:			
Minimum	2.5A @ 5000 rpm	1.8A @ 2500 rpm	–
Maximum	4.0A @ 8000 rpm	4.0A @ 8000 rpm	–
Lighting output:			
Minimum	7.0V @ 2500 rpm	7.0A @ 2500 rpm	–
Maximum	9.0V @ 8000 rpm	9.0A @ 8000 rpm	–
Wattage	–	–	47W @ 5000 rpm
Battery			
Capacity	6V 4Ah	6V 4Ah	–
Regulator/rectifier	Sealed, solid state	Sealed, solid state	–

Bulb wattages – US (UK)*	XL models	XR models
Headlamp	36.5/35 W (35/35 W)	25/25W (25/25W)
Tail/stop lamp	3/32 CP (5/21W)	2CP (3W)
Parking lamp (UK only)	4W	–
Indicators	21 CP (21W)	–
Speedometer illumination	1 CP (1.7W)	2 CP (3W)
Neutral warning	2 CP (3.0W)	–
Indicator warning	1 CP (1.7W)	–
High beam warning	1 CP (1.7W)	–

*Note: Bulb wattages may have been changed by dealer on UK versions of XR models. XR models have no provision for a brake light, indicators or high beam warning lamp as standard. This may have been modified on UK machines.

1 General description

The Honda XL and XR models covered by this manual employ two distinctly different electrical systems. The XL versions utilise a similar electrical system to those fitted to most lightweight road machines. Power from the alternator charging coil is fed to the regulator/rectifier unit mounted beneath the fuel tank, where the alternating current (ac) output is converted to direct current (dc) and the charge rate adjusted to match the demands of the lighting system. This system caters for the indicators, horn, brake lamp and parking lamps, where these are fitted. The headlamp circuit operates separately, and is fed directly from the alternator. It follows that the headlamp will work only when the engine is running.

The system fitted to the XR models is a simple direct lighting arrangement in which no battery is fitted. It is a rather basic lighting system intended to make the machine legal for use on public road sections of an enduro course, rather than to provide the more sophisticated facilities found on road or trail machines.

In the case of XR models sold in the UK, it is unlikely that some form of modification will have been carried out to equip them with a horn and a brake light and switch, these being required by law for use on public roads. These are not fitted as standard by the factory, and any such modification will have been carried out and devised by the dealer that sold the machine. It follows that details of these fittings are not available.

2   Checking the electrical system: general

1   Many of the test procedures applicable to motorcycle electrical systems require the use of test equipment of the multimeter type. Although the test themselves are quite straightforward, there is a real danger, particularly on alternator systems, of damaging certain components if wrong connections are made. It is recommended, therefore, that no attempt be made to investigate faults in the charging system, unless the owner is reasonably experienced in the field. A qualified Honda Service Agent will have in his possession the necessary diagnostic equipment to effect an economical repair.
2   Owners with multimeters will no doubt be aware of their value in checking circuit continuity. This is done with the meter set on the resistance or Ohms scale, and can be used to check leads, switch terminals connectors and insulation as required. An inexpensive alternative can be made up using a dry battery and a small bulb, as shown in Fig.3.2, page 95. In all but a few tests where specific resistance readings are required, this arrangement will be perfectly adequate.

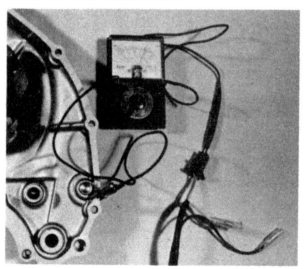

2.2 Multimeter is invaluable when checking system

3   Checking the electrical system: XR models

1   The XR electrical system on the XR models is unlikely to cause many problems in view of the limited number of components whch can fail. Most likely of the few possible faults are blown bulbs, damaged or shorted wiring or a faulty switch unit.
2   In the event of a total failure, check through the system, starting at the lighting coil. This can be checked by measuring the resistance between the white/yellow lead and earth (ground). Trace the output leads back to the connector beneath the fuel tank, separating it to expose the various terminals. Using a multimeter set on the resistance scale (Ohms), check that a continuity reading is indicated when the probe leads are connected as described above. If so, the lighting coil can be considered functional, and attention can be turned elsewhere.
3   If the alternator is working properly, the fault will lie in the

wiring to the lighting switch or in the switch unit itself. Trace the wiring back to the switch, using the wiring diagram at the end of the book as guidance. Look for signs of breakage or damage and for corroded or waterlogged connectors or terminals.
4   The operation of the switch can be checked by measuring continuity between the various leads with the switch set at different positions. This should be done at the connectors behind the headlamp unit, which should be removed to permit easy access. The appropriate tests and correct readings are shown in the table below.

LIGHTING SWITCH	DIMMER SWITCH	WHITE/ YELLOW	BROWN	BLUE	WHITE
OFF					
ON	Hi	o————	———o————	———o	
	(N)	o————	———o————	———o————	———o
	Lo	o————	———o		———o

5   Partial failure of the electrical system can generally be attributed to a faulty connection or a blown bulb. Check the bulb first, ensuring thaat it is sound by measuring for continuity across its filament or by substitution. If the bulb is intact, look for a broken or shorted lead between the relevant switch terminal and the bulb. Check any connectors and terminals for corrosion or mechanical failure and ensure that the switch is functioning normally as described above.

4   Checking the charging system: XL models

1   Before any test on the charging system is made, it is important that the battery is checked. It must be in sound condition and fully charged if an accurate diagnosis is to be obtained. Very often it will be found that the battery itself is responsible for the fault, and this check is therefore most important. For details, reference should be made to Section 7 of this Chapter. In addition to this, a careful examination of all wiring and connectors should be made to eliminate these as a possible cause of the problem.
2   The test requires the use of a voltmeter with a range of 0 – 20 volts dc, and an ammeter capable of reading at least ‡6.0 amps. The meters should be connected into the electrical system as, shown in the accompanying diagram. The voltmeter is connected across the battery leads, whilst the ammeter is inserted in the red lead between the battery and the regulator/rectifier unit. The engine should be at normal operating temperature during the tests. Check the meter readings at the prescribed engine speeds and compare them with the figures given below. Note that to carry out the test accurately, a tachometer will be required, but a good indication of the conditions system may be gleaned without precise figures.

Charging system output test

XL 250S	*Lights off*	*Lights on (main beam)*
Initial charge starts at	*800 rpm*	*1000 rpm*
At 5000 rpm	*4.0A minimum/ 8.0V*	*2.6A minimum/ 7.5V*
At 8000 rpm	*5.5A maximum/ 8.9V*	*4.0A maximum/ 8.0V*

XL 500S	*Lights off*	*Lights on (main beam)*
Initial charge starts at	*500 rpm*	*1200 rpm*
At 2500 rpm	*3.2A minimum/ 8.0V*	*1.8A minimum/ 7.5V*
At 8000 rpm	*5.5A maximum/ 8.9V*	*4.0A maximum/ 8.0V*

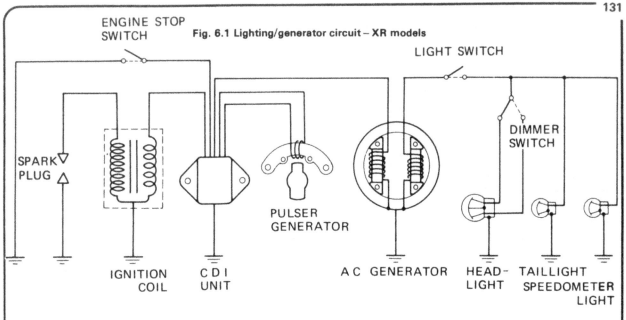

Fig. 6.1 Lighting/generator circuit – XR models

ENGINE STOP SWITCH

LIGHT SWITCH

DIMMER SWITCH

SPARK PLUG

PULSER GENERATOR

IGNITION COIL

C D I UNIT

A C GENERATOR

HEAD-LIGHT

TAILLIGHT SPEEDOMETER LIGHT

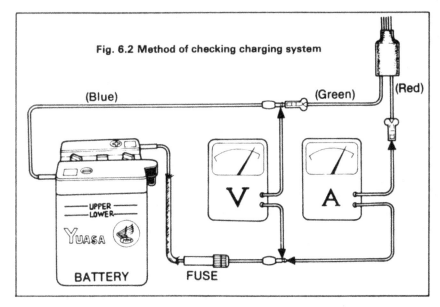

Fig. 6.2 Method of checking charging system

(Blue)

(Green)

(Red)

UPPER LOWER

YUASA

BATTERY

FUSE

Fig. 6.3 Lighting/generator circuit – XL model

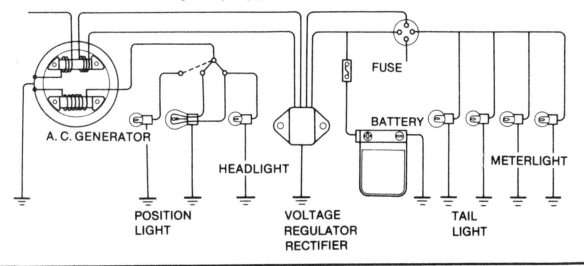

A. C. GENERATOR

HEADLIGHT

FUSE

BATTERY

METERLIGHT

POSITION LIGHT

VOLTAGE REGULATOR RECTIFIER

TAIL LIGHT

5 Alternator: resistance tests

1   If the output tests described in Section 4 have indicated that the alternator is deficient, it will be necessary to check the resistance of the charging coil and lighting coil windings using a multimeter. The test connections are made at the connector block beneath the fuel tank, and the latter should be removed to gain access to it. No further dismantling is required.
2   Separate the connector and measure the charging coil resistance by connecting the meter probe leads to the pink and the yellow output lead terminals. The lighting coil is tested in the same manner, using the white/yellow lead and earth as probe connections. In both cases, a reading of 0.2 – 1.0 Ohm should be obtained.
3   If the above tests indicate that the alternator is working correctly, the fault is likely to lie with the regulator/rectifier unit. See Section 6 for details.

6 Regulator/rectifier unit: testing

1   The regulator/rectifier unit is tested by measuring the resistance between the various pairs of leads. The test sequence is perfectly straightforward, but requires the use of specific electrical testers. The readings obtained from each type differ, so it is essential that one type or the other of the

recommended models is used. No general resistance figures are available. For all practical purposes, this will mean that the owner must rely on a Honda Service Agent to perform the test, but for those having access to the SANWA ELECTRICAL TESTER, part number 07308-0020000, or a KOWA ELECTRICAL TESTER (TH-5H), the test connections and expected results are given in the tables below.

SANWA: [kΩ]

+ Probe / − Probe	Yellow	Pink	Red	Black	Green
Yellow		∞	1–25	∞	∞
Pink	10–250		1–25	1.5–35	6–150
Red	∞	∞		∞	∞
Black	5–120	5–120	10–250		5–45
Green	1–25	1–25	3–80	1.5–40	

KOWA: [x 100]

+ Probe / − Probe	Yellow	Pink	Red	Black	Green
Yellow		∞	2–50	∞	∞
Pink	100–∞		2–50	6–150	50–∞
Red	∞	∞		∞	∞
Black	50–∞	50–∞	100–∞		50–500
Green	2–50	2–50	8–200	8–200	

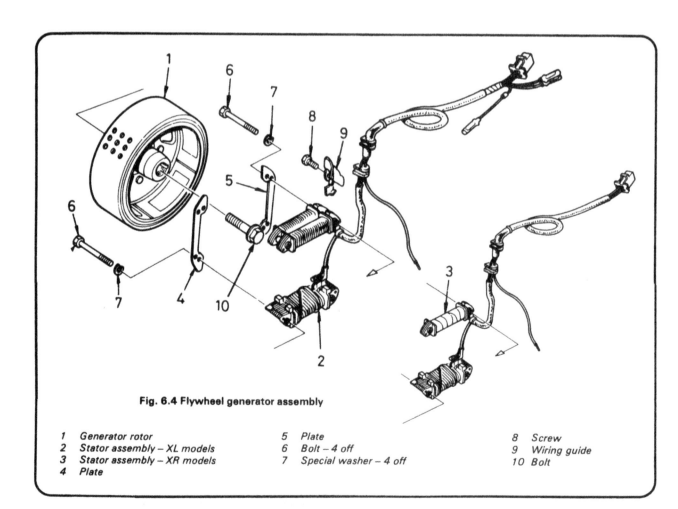

Fig. 6.4 Flywheel generator assembly

1	Generator rotor	5	Plate	8	Screw
2	Stator assembly – XL models	6	Bolt – 4 off	9	Wiring guide
3	Stator assembly – XR models	7	Special washer – 4 off	10	Bolt
4	Plate				

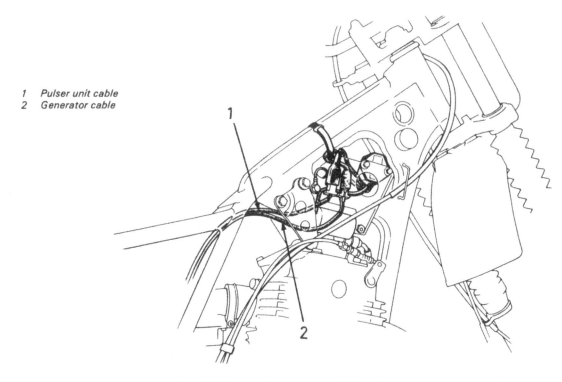

1   Pulser unit cable
2   Generator cable

Fig. 6.5 Pulser unit alternator cable locations

7   Battery: charging procedure – XL models

1   Whilst the machine is used on the road it is unlikely that the battery will require attention other than routine maintenance because the generator will keep it fully charged. However, if the machine is used for a succession of short journeys only, mainly during the hours of darkness when the lights are in full use, it is possible that the output from the generator may fail to keep pace with the heavy electrical demand, especially if the machine is parked with the lights switched on. Under these circumstances it will be necessary to remove the batteery from time to time to have it charged independently.

2   The battery is mounted in a plastic holder on the right-hand side of the machine, to the rear of the side panel. It can be removed by releasing the cover and removing the nut which secures the holding strap. Disconnect the battery leads and lift the battery out of its holder.

3   The battery fitted to the XL 250S and the XL500S is rated at 6 volts, 4Ah (ampere-hours). If it is necessary to give the battery an external charge, the charger must be set at 6 volts and the charge rate should not exceed 1.2 amps. A higher charge rate should be avoided as there is a real risk of damage due to buckled plates caused by overheating. If at all possible, use a current-controlled charger set at the lowest practicable charge rate. A rate 0.5 amps for 8 hours will charge a flat battery in safety.

4   Before charging, remove each cell cap to prevent gas pressure building up. Batteries can, and do, explode if this precaution is overlooked. During charging, the electrolyte will give off bubbles of hydrogen and oxygen. The former is highly inflammable and when mixed with the latter, is explosive. For this reason, keep the battery well away from naked lights or heat sources.

5   Connect the charger, observing polarity, and commence charging. If at any time the battery becomes hot (specifically, higher than 45°C [117°F]), stop charging at once and allow the

battery to cool before resuming. At all times, take great care to avoid acid splashes on skin, eyes or clothing.

6   When the battery is removed from the machine, clean the battery top. If the terminals are corroded scrape them clean and cover them with Vaseline (not grease) to protect them from further attack. If a vent tube is fitted, make sure it is not obstructed and that it is arranged so that it will not discharge over any parts of the machine.

7   If the machine is laid up for any period of time, the battery should be removed and given a 'refresher' charge every six weeks or so, in order to maintain it in good condition.

7.2 Battery is housed beneath plastic cover

8 Headlamp: adjustment and bulb replacement

XL models

1 The XL models are equipped with a conventional round headlamp unit fitted with either a separate bulb and reflector unit or a sealed beam unit, depending on the country to which the machine is exported. In the case of the UK and most European countries, the former type is fitted, and adjustment is limited to vertical movement. In the case of US and General Market models where a sealed beam unit is fitted, additional horizontal adjustment is provided by a small screw in the front edge of the headlamp rim.

2 The vertical adjustment is carried out by slackening the two headlamp mounting bolts. This will allow the headlamp assembly to be pivoted around the mounting bolt axis and set at the prescribed position. When adjustment has been completed, the mounting bolts should be tightened to retain the setting.

3 In the UK, regulations stipulate that, when on dipped beam, the headlamp must be arranged so that the light will not dazzle a person at a distance of 25 feet, and whose eye level is 3 feet 6 inches above that plane. In most other countries, similar regulations apply, but the specific requirements should be checked locally. This setting can be approximated by making a mark on a flat wall, 3 feet 6 inches (1.067 metres) from the road surface. Place the machine 25 feet (7.62 metres) away from the wall and set the lamp so that the light on dipped beam does not fall above the mark. The setting should be carried out with the machine off its stand, and with the rider (and passenger, if carried regularly), seated normally. Similar requirements exist in many other countries, and these should be investigated prior to adjustment.

4 Horizontal adjustment is effected by turning the small screw in the headlamp rim. This can be found at the 8 o'clock position when viewed from the front. Turning the screw clockwise will move the beam to the left. The setting of the horizontal alignment is to some extent discretionary. It should, however, be approximately straight ahead when viewed from the machine.

5 As already mentioned, a bulb or sealed beam unit may be fitted, depending on the model and country in which it was sold. Replacement details are similar for both models. In the event of failure, release the headlamp unit from the shell after removing the retaining screws. On models fitted with the bulb-type unit, disengage the bulbholder to release the bulb. On machines with a sealed-beam unit, it will be necessary to dismantle the horizontal adjustment screw and to separate the headlamp rim from the unit by removing the two pivot screws. Note carefully the order in which the assembly is dismantled to ensure correct reassembly. The sealed beam unit will be correctly aligned if the adjustment holes in the rim and unit are adjacent to each other.

6 It is essential that the headlamp vertical and horizontal adjustments are checked after reassembly, irrespective of which type of unit it fitted.

7 On UK models a pilot lamp is incorporated in the headlamp reflector. The bulbholder is a push fit in the reflector, and houses a simple bayonet cap bulb.

XR models

8 The XR models make use of a small rectangular unit mounted in the plastic number plate on the fork yokes. The unit is retained at the top by a split pin which passes through holes in the mounting bracket and allows the assembly to pivot. Adjustment is provided by a spring-loaded screw immediately below the unit.

9 Access to the headlamp assembly is gained by removing the plastic number plate. Ths is secured by two bolts which pass through lugs at the top of the plate and by a third bolt which passes through the lower edge of the plate. The headlamp unit can be released by dismantling the adjuster and withdrawing the split pin at the upper edge.

10 All models use a bulb-type unit. The bulb is retained by a three-pronged bayonet holder which is in turn covered by a rubber shroud.

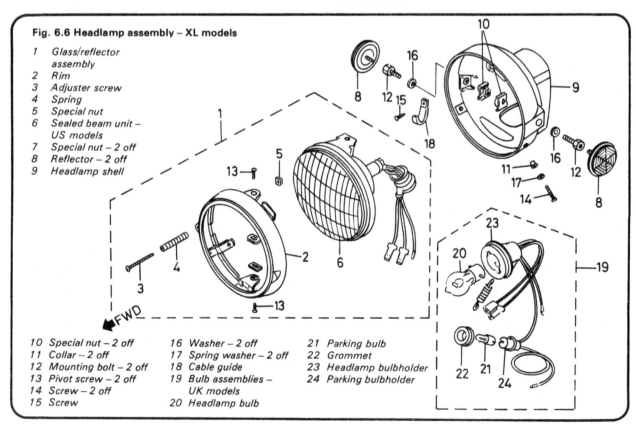

Fig. 6.6 Headlamp assembly – XL models

1   Glass/reflector assembly
2   Rim
3   Adjuster screw
4   Spring
5   Special nut
6   Sealed beam unit – US models
7   Special nut – 2 off
8   Reflector – 2 off
9   Headlamp shell

10 Special nut – 2 off
11 Collar – 2 off
12 Mounting bolt – 2 off
13 Pivot screw – 2 off
14 Screw – 2 off
15 Screw

16 Washer – 2 off
17 Spring washer – 2 off
18 Cable guide
19 Bulb assemblies – UK models
20 Headlamp bulb

21 Parking bulb
22 Grommet
23 Headlamp bulbholder
24 Parking bulbholder

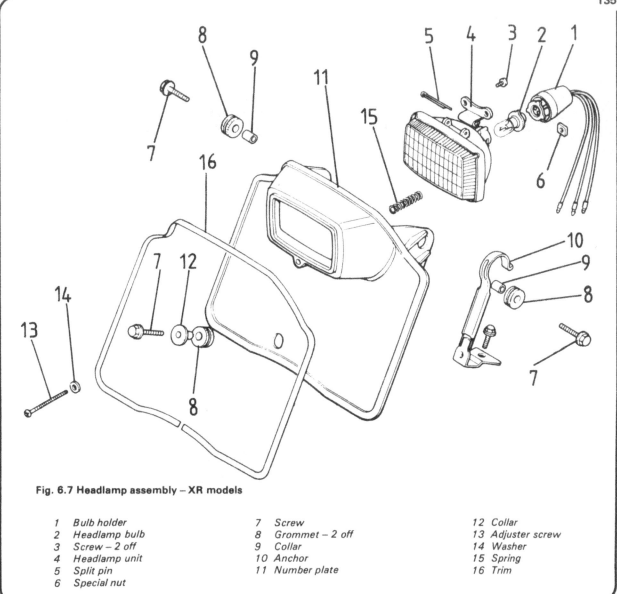

Fig. 6.7 Headlamp assembly – XR models

1	Bulb holder	7	Screw	12	Collar
2	Headlamp bulb	8	Grommet – 2 off	13	Adjuster screw
3	Screw – 2 off	9	Collar	14	Washer
4	Headlamp unit	10	Anchor	15	Spring
5	Split pin	11	Number plate	16	Trim
6	Special nut				

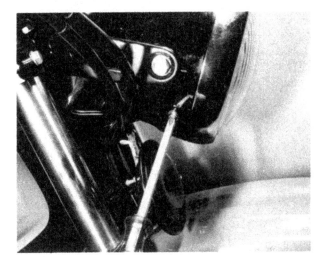

8.5a Remove screws to free headlamp assembly

8.5b Disconnect spring to release bulbholder

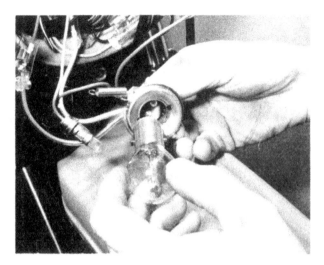

8.5c Bulb is bayonet fitting in holder

8.7 Pilot lamp (UK versions) is push-fit in reflector

9   Rear lamp assembly: bulb renewal

1   The rear lamp lens can be removed after its three retaining
screws have been released (two screws, XR models). The XL
versions are fitted with a twin-filament bulb, one of which
serves as a tail lamp and also illuminates the number (license)
plate. The remaining filament is operated by the brake light
switch.
2   In the case of the XR versions, a small single-filament bulb
is fitted, functioning as a tail and number plate light only. In
either case, the bulb is a bayonet fitting, that of the XL having
offset pins to ensure that it is installed correctly.
3   When refitting the lens, check that the seal surface is sound
and clean to prevent the ingress of dirt or water. If the bulb
tends to blow frequently, suspect vibration of the mudguard, or
if the bulb is blackened, check the wiring and earth connections.

10 Flashing indicator lamps: XL models only

1   The indicator lamps are attached to the underside of the
handlebars, and to the frame immediately to the rear of the dual
seat. The lamps are connected to the machine via rubber links
which reduces the risk of breakage should the machine be
dropped, the front units having additional protection in being
masked by the handlebar assembly.
2   Access to the bulb is gained by removing the two lens
securing screws. The bulbs are of the conventional bayonet cap
type and have single filaments.

11 Flasher unit: location and replacement – XL models

1   The flasher relay unit is located under the fuel tank, being
rubber-mounted to the frame top tube.
2   If the flasher unit is functioning correctly, a series of audible
clicks will be heard when the indicator lamps are in action. If the
unit malfunctions and all the bulbs are in working order, the
usual symptom is one initial flash before the unit goes dead; it
will be necessary to replace the unit complete if the fault cannot
be attributed to any other cause.
3   Take great care when handling the unit because it is easily
damaged if dropped.

12 Speedometer head: replacement of bulbs

1   The speedometer illumination bulb (all models) and warning

lamp bulbs (XL models only) are retained in the underside of the
speedometer or instrument panel by push-fit rubber holders.
These are simply prised from their locating holes and the
bayonet cap bulb disengaged.

13 Stop lamp switches: location and replacement – XL models

1   Two stop lamp switches are fitted to the machine, which
work independently of one another, depending on which brake
is operated.
2   The front brake switch is fitted to the handlebar lever stock
and is a mechanical push-off type, being operated when the
lever is removed. The switch is a push fit in the housing boss,
and is detached by depressing a small pin in the underside with
a piece of wire or a small screwdriver.
3   The rear brake switch is mounted on the frame on the right-
hand side, above the rear brake pedal. It can be adjusted by
means of a locknut, and should be set so that the light comes
on as soon as the pedal is depressed. This is especially
important when the rear brake has been readjusted.

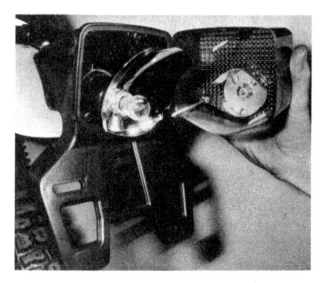

9.1 Rear lamp lens (XL models) is retained by three screws

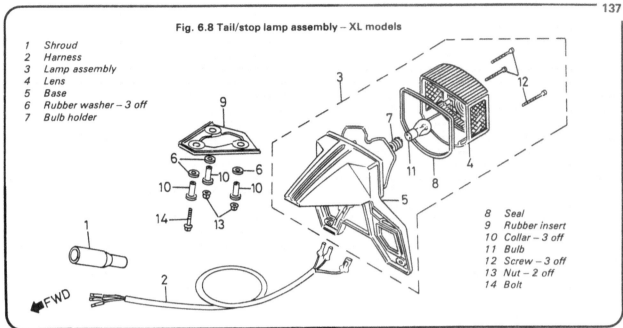

Fig. 6.8 Tail/stop lamp assembly – XL models

1 Shroud
2 Harness
3 Lamp assembly
4 Lens
5 Base
6 Rubber washer – 3 off
7 Bulb holder

8 Seal
9 Rubber insert
10 Collar – 3 off
11 Bulb
12 Screw – 3 off
13 Nut – 2 off
14 Bolt

◀FWD

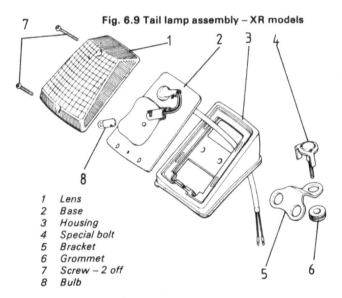

Fig. 6.9 Tail lamp assembly – XR models

1 Lens
2 Base
3 Housing
4 Special bolt
5 Bracket
6 Grommet
7 Screw – 2 off
8 Bulb

10.1 Remove indicator lens to gain access to bulb (XL models)

11.1 Indicator unit (arrowed) is mounted beneath fuel tank

12.1a Instrument lamp bulbs are a push fit

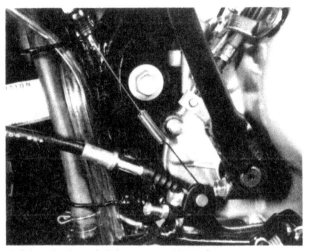

12.1b Instrument panel is secured by studs (arrowed)

13.3 Rear brake switch is connected by spring

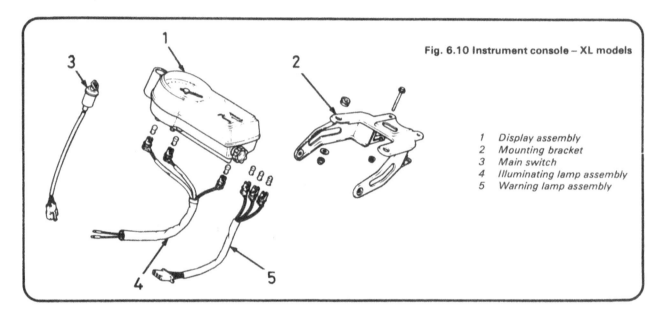

Fig. 6.10 Instrument console – XL models

1   Display assembly
2   Mounting bracket
3   Main switch
4   Illuminating lamp assembly
5   Warning lamp assembly

14 Horn: location and examination – XL models

1   The horn is located on the lower fork yoke, secured to it by a flexible steel strip which isolates it from vibration and shocks. It should not require frequent attention, but should be kept clean, and the terminals checked occasionally for tightness. If the horn fails to operate, check that current is reaching it by inserting a multimeter set on 20 v dc or a suitable bulb between the horn leads. With the ignition switched on and the horn button depressed, the bulb should light or the meter needle deflect to indicate that the circuit is intact. If this is not the case, trace the wiring, connectors and fuse until the fault is identified.

2   After an extended period of use, the horn's internal contacts may become so eroded that it will cease to operate. If this condition arises, experiment by turning the contact adjustment screw a fraction of a turn at a time until the horn operates. If this succeeds in restoring the horn to working order, continue adjustment to obtain a clear, loud note. Further maintenance is impracticable, and as a last resort a new unit must be fitted. Remember that in the UK and most other countries, a horn in working order is a statutory requirement on all machines.

15 Wiring: layout and examination

1   The wiring harness is colour-coded and will correspond with the accompanying wiring diagram. Where socket connectors are used, they are designed so that reconnection can be made in the correct position only.

2   Visual inspection will show whether there are any breaks or frayed outer coverings which will give rise to short circuits. Another source of trouble may be the snap connectors and sockets, where the connector has not been pushed fully home in the outer housing.

3   Intermittent short circuits can often be traced to a chafed wire that passes through or is close to a metal component such as a frame member. Avoid tight bends in the lead or situations where a lead can become trapped between casings.

16 Switches: maintenance and fault finding

1   The XL models are equipped with a key-operated main switch, or ignition switch, which controls the feed from the battery to the electrical system and also serves to earth the

ignition system when switched off. A separate ignition kill switch on the right-hand handlebar end duplicates the latter function and is combined with a lighting switch. The lighting switch has three positions, Off, Park and Headlamp. A second switch assembly at the left-hand end of the handlebar contains the dipswitch, indicator switch and horn switch.

2   The XR models have a much simplified system. There is no key-operated main switch, the kill switch controlling the operation of the ignition system. A switch unit on the left-hand end of the handlebar operates the lighting system, which functions only when the engine is running.

3   The switches are generally reliable, but may give trouble if they become soaked in water or corroded. One of the multi-purpose maintenance sprays, such as WD40, can be used to good effect if cleaning proves necessary. Generally speaking, it is difficult to repair a faulty unit, the normal solution being renewal of the assembly concerned. If, however, the switch is imperative, it may be worthwhile attempting to dismantle and clean or repair the faulty contacts. There is nothing to lose if the attempt fails.

4   Switch operation can be checked by using a multimeter to establish whether continuity exists between the appropriate terminals. Measurement is made by connecting the meter probe leads at the switch connector block, there being no need to dismantle the switches themselves. The tables that will be found in the appropriate wiring diagram at the end of this Chapter show the various switch positions and the terminals which should be connected or isolated in each.

17 Fuse: location and renewal

1   A single 10 amp fuse is fitted to the positive (+) battery lead on the XL models. It is contained in a small cylindrical line fuse holder. No fuse is fitted to the XR versions.

2   If a fuse blows, it should not be renewed until a check has shown whether a short circuit has occurred. This will involve checking the electrical circuit to identify and correct the fault. If this precaution is not observed, the replacement fuse, which may be the only spare, may blow immediately on connection.

3   When a fuse blows whilst the machine is running and no spare is available, a 'get you home' remedy is to remove the blown fuse and wrap it in silver paper before replacing it in the fuse holder. The silver paper will restore electrical continuity by bridging the broken wire within the fuse. This expedient should never be used if there is evidence of a short circuit or other major electrical fault, otherwise more serious damage will be caused. Renew the 'doctored' fuse at the earliest possible opportunity to restore full circuit protection.

15.1 Wiring is colour-coded to aid tracing

16.3 Switch halves can be separated for cleaning

17.1 Fuse is contained in-line holder in battery lead

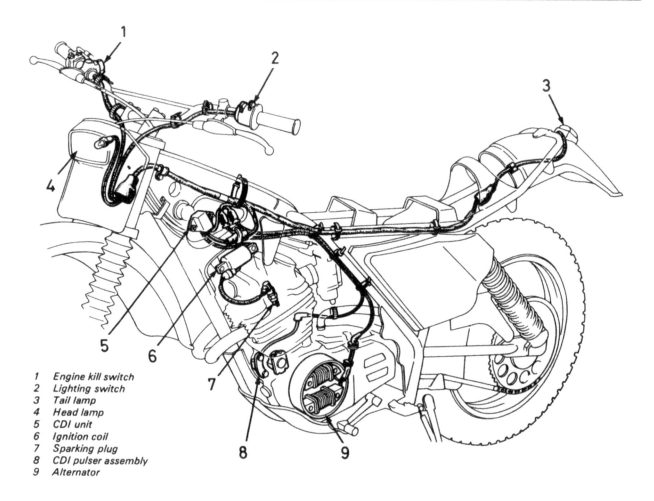

1 Engine kill switch
2 Lighting switch
3 Tail lamp
4 Head lamp
5 CDI unit
6 Ignition coil
7 Sparking plug
8 CDI pulser assembly
9 Alternator

Fig. 6.11 Location of electrical components – XR models

18 Fault diagnosis: electrical system

Symptom	Cause	Remedy
Complete electrical failure (XL models)	Blown fuse	Check wiring and electrical components for short circuit before fitting new fuse.
	Isolated battery	Check battery connections, also whether connections show signs of corrosion.
Dim lights, horn and indicators (XL models)	Discharged battery	Remove battery and charge with battery charger. Check generator output and voltage regulator settings.
Constantly blowing bulbs	Vibration or poor earth connection	Check security of bulb holders. Check earth return connections.
Parking lights dim rapidly (XL models)	Battery will not hold charge	Renew battery at earliest opportunity
Flashing indicators do not operate (XL models)	Blown bulb
Damaged flasher unit	Renew bulb.
Renew flasher unit.

The XL250 R-C model

Engine/gearbox unit

Chapter 7 The 1981 to 1984 models

Contents

Specifications

Note: *Specifications are given below where they differ from those shown in Chapters 1 to 6, and subject to availability at the time of writing. For specifications covering the 1981-82 XR250 and 500 models refer to the US versions shown below.*

Specifications relating to Chapter 1 – UK XL models

Engine

Compression ratio – XL250 R-C	9.3:1
Power output:	
XL250 S-B	15.0 kW (20.4 PS) @ 7500 rpm
XL250 R-C	16.1 kW (22.0 PS) @ 7500 rpm
XL500 R-C	24.3 kW (33.0 PS) @ 6500 rpm
Maximum torque:	
XL250 S-B	20.0 Nm (2.04 kg/m) @ 6000 rpm
XL250 R-C	21.0 Nm (2.10 kg/m) @ 7000 rpm
XL500 R-C	39.0 Nm (3.90 kg/m/28.2 ft lb) @ 5000 rpm

Compression pressure
At cranking speed XL250 R-C .... 14.0 kg cm² (199 psi)

Balancer assembly
Type XL250 R-C .... Single gear-driven front balancer shaft

Clutch
Clutch spring free length:
    XL500 R-C .... 44.1 mm (1.74 in)
    Service limit .... 42.5 mm (1.67 in)
Clutch spring preload/length XL500 R-C .... 23.7 − 26.3 kg @ 27.0 mm (52.2 − 58.0 lb @ 1.06 in)

Gearbox
Type − XL250 R-C .... Six-speed constant mesh
Ratios − XL250 R-C:
    1st 2.307:1 (43/13T)
    2nd 2.111:1 (38/18T)
    3rd 1.590:1 (35/22T)
    4th 1.280:1 (32/25T)
    5th 1.074:1 (29/27T)
    6th 0.931:1 (27/29T)
Gear pinion internal diameters − XL250 R-C:
    Output shaft 1st, 4th .... 25.020 − 25.041 mm (0.9850 − 0.9859 in)
    Service limit .... 25.10 mm (0.988 in)
    Output shaft 2nd .... 27.020 − 27.053 mm (1.0638 − 1.0651 in)
    Service limit .... 27.12 mm (1.068 in)
    Output shaft 3rd, input shaft 6th .... 28.020 − 28.053 mm (1.1032 − 1.1045 in)
    Service limit .... 28.12 mm (1.107 in)
    Input shaft 5th .... 28.020 − 28.041 mm (1.1032 − 1.1040 in)
    Service limit .... 28.10 mm (1.106 in)
Gear pinion bush outside diameters − XL250 R-C:
    Output shaft 3rd, input shaft 6th .... 27.969 − 27.980 mm (1.1011 − 1.1016 in)
    Service limit .... 27.90 mm (1.098 in)
Selector fork ID − XL250 R-C:
    Centre .... 12.000 − 12.018 mm (0.4724 − 0.4732 in)
    Service limit .... 12.05 mm (0.474 in)
    Right and left .... 15.000 − 15.018 mm (0.5906 − 0.5913 in)
    Service limit .... 15.05 mm (0.593 in)
Selector fork claw thickness:
    XL250 R-C .... 4.98 − 5.00 mm (0.196 − 0.197 in)
    Service limit .... 4.5 mm (0.18 in)

Secondary transmission
Reduction ratio:
    XL250 R-C 3.143:1 (44/14T)
    XL500 R-C 2.733:1 (41/15T)

Torque settings

Component	XL250 R-C		XL500 R-C	
	kgf m	lbf ft	kgf m	lbf ft
Kickstart stopper plate	N/Av	N/Av	1.8 − 2.5	13 − 18
Kickstart spring anchor pin	N/Av	N/Av	2.2 − 3.0	16 − 22
Gear selector drum retainer	0.9 − 1.3	7 − 9	0.9 − 1.3	7 − 9
Upper crankcase bolts:				
6 mm	1.0 − 1.4	7 − 10	1.0 − 1.4	7 − 10
8 mm	2.0 − 2.6	14 − 19	2.2 − 2.8	16 − 20
Lower crankcase bolts:				
6 mm	1.0 − 1.4	7 − 10	1.0 − 1.4	7 − 10
8 mm	2.2 − 2.8	16 − 20	N/App	N/App
9 mm	N/App	N/App	2.7 − 3.3	20 − 24
10 mm	N/App	N/App	3.2 − 3.8	23 − 27
Balancer holder lock bolt	2.0 − 2.6	14 − 19	2.2 − 2.8	16 − 20
Cam chain tensioner	0.8 − 1.2	6 − 9	1.0 − 1.4	7 − 10
Primary drive pinion nut	4.5 − 6.0	33 − 43	4.5 − 6.0	33 − 43
Clutch centre nut	4.5 − 6.0	33 − 43	4.5 − 6.0	33 − 43
Alternator rotor bolt	10.0 − 12.0	72 − 87	10.0 − 12.0	72 − 87
Cylinder head cover bolts:				
6 mm (plain)	1.0 − 1.4	7 − 10	1.0 − 1.4	7 − 10
7 mm (plain)	1.0 − 1.4	7 − 10	1.3 − 1.7	9 − 12
6 mm (with sealing washer)	1.0 − 1.2	7 − 9	N/App	N/App
Cylinder head bolts 8 mm	3.5 − 4.0	25 − 29	N/App	N/App
Cylinder head cap nut	N/App	N/App	2.2 − 2.8	16 − 20

Torque settings (continued)

Component	XL250 R-C kgf m	lbf ft	XL500 R-C kgf m	lbf ft
Cylinder 6 mm bolt	0.8 – 1.2	6 – 9	1.0 – 1.4	7 – 10
Cylinder 8 mm nut	2.2 – 2.8	16 – 20	2.2 – 2.8	16 – 20
Cam sprocket bolt	1.7 – 2.3	12 – 17	1.7 – 2.3	12 – 17
Valve adjuster locknut	0.8 – 1.2	6 – 9	1.5 – 1.8	11 – 13
Alternator stator bolt	0.8 – 1.2	6 – 9	N/App	N/App
Alternator stator base	0.9 – 1.3	7 – 9	N/App	N/App
Valve adjuster inspection cover	0.8 – 1.2	6 – 9	1.0 – 1.4	7 – 10
Carburettor mounting clip	0.3 – 0.5	2 – 4	0.3 – 0.5	2 – 4
Decompressor cable locknut	N/App	N/App	0.5 – 0.7	4 – 5
Selector drum stopper plate	1.0 – 1.4	7 – 10	N/App	N/App
Selector cam bolt	1.0 – 1.4	7 – 10	N/App	N/App
Gearbox sprocket	0.8 – 1.2	6 – 9	N/App	N/App
Engine oil drain plug	2.0 – 3.0	14 – 22	3.0 – 4.0	22 – 29
Spark plug	1.5 – 2.0	11 – 14	1.5 – 2.0	11 – 14

Specifications relating to Chapter 1 – US XL models

Engine

Compression ratio – XL250 R 1982	9.3:1
Power output:	
XL250 R 1982	22.0 bhp ⊕ 8000 rpm
XL500 R 1982	32.4 bhp ⊕ 6500 rpm
Maximum torque:	
XL250 R 1982	2.17 kg/m (15.7 ft lb) ⊕ 6500 rpm
XL500 R 1982	4.00 kg/m (28.2 ft lb) ⊕ 5000 rpm

Compression pressure

At cranking speed – XL250 R 1982	14.0 kg/cm^2 (199 psi)

Balancer assembly

Type XL250 R 1982	Single gear-driven front balancer shaft

Gearbox

Type – XL250 R 1982	Six-speed constant mesh
Ratios – XL250 R 1982:	
1st	2.307:1 (43/13T)
2nd	2.111:1 (38/18T)
3rd	1.590:1 (35/22T)
4th	1.280:1 (32/25T)
5th	1.074:1 (29/27T)
6th	0.931:1 (27/29T)
Gear pinion internal diameters – XL250 R 1982:	
Output shaft 1st, 4th	25.020 – 25.041 mm (0.9850 – 0.9859 in)
Service limit	25.10 mm (0.988 in)
Output shaft 2nd	27.020 – 27.053 mm (1.0638 – 1.0651 in)
Service limit	27.12 mm (1.068 in)
Output shaft 3rd, input shaft 6th	28.020 – 28.053 mm (1.1032 – 1.1045 in)
Service limit	28.12 mm (1.107 in)
Input shaft 5th	28.020 – 28.041 mm (1.1032 – 1.1040 in)
Service limit	28.10 mm (1.106 in)
Gear pinion bush outside diameters – XL250 R 1982:	
Output shaft 3rd, input shaft 6th	27.969 – 27.980 mm (1.1011 – 1.1016 in)
Service limit	27.90 mm (1.098 in)
Selector fork ID – XL250 R 1982:	
Centre	12.000 – 12.018 mm (0.4724 – 0.4732 in)
Service limit	12.05 mm (0.474 in)
Right and left	15.000 – 15.018 mm (0.5906 – 0.5913 in)
Service limit	15.05 mm (0.593 in)
Selector fork claw thickness:	
XL250 R 1982	4.98 – 5.00 mm (0.196 – 0.197 in)
Service limit	4.5 mm (0.18 in)

Secondary transmission

Reduction ratio:	
XL250 R 1982	3.357:1 (47/14)
XL500 S 1981	2.929:1 (41/14)
XL500 R 1982	2.867:1 (43/15)

Decompressor mechanism

Starter decompressor cable free play – XL500 S 1981 .... 1 – 2 mm ($\frac{1}{16}$ in)
Manual decompressor cable free play – XL500 S 1981 ...... 5 – 8 mm ($\frac{3}{16}$ – $\frac{5}{16}$ in)

Torque settings

Component	XL250 R 1982		XL500 R 1982	
	kgf m	lbf ft	kgf m	lbf ft
Kickstart stopper plate	N/Av	N/Av	1.8 – 2.5	13 – 18
Kickstart spring anchor pin	N/Av	N/Av	2.2 – 3.0	16 – 22
Gear selector drum retainer	0.9 – 1.3	7 – 9	0.9 – 1.3	7 – 9
Upper crankcase bolts:				
6 mm	1.0 – 1.4	7 – 10	1.0 – 1.4	7 – 10
8 mm	2.0 – 2.6	14 – 19	2.2 – 2.8	16 – 20
Lower crankcase bolts:				
6 mm	1.0 – 1.4	7 – 10	1.0 – 1.4	7 – 10
8 mm	2.2 – 2.8	16 – 20	N/App	N/App
9 mm	N/App	N/App	2.7 – 3.3	20 – 24
10 mm	N/App	N/App	3.2 – 3.8	23 – 27
Balancer holder lock bolt	2.0 – 2.6	14 – 19	2.2 – 2.8	16 – 20
Cam chain tensioner	0.8 – 1.2	6 – 9	1.0 – 1.4	7 – 10
Primary drive pinion nut	4.5 – 6.0	33 – 43	4.5 – 6.0	33 – 43
Clutch centre nut	4.5 – 6.0	33 – 43	4.5 – 6.0	33 – 43
Alternator rotor bolt	10.0 – 12.0	72 – 87	10.0 – 12.0	72 – 87
Cylinder head cover bolts:				
6 mm (plain)	1.0 – 1.4	7 – 10	1.0 – 1.4	7 – 10
7 mm (plain)	1.0 – 1.4	7 – 10	1.3 – 1.7	9 – 12
6 mm (with sealing washer)	1.0 – 1.2	7 – 9	N/App	N/App
Cylinder head bolts – 8 mm	3.5 – 4.0	25 – 29	N/App	N/App
Cylinder head cap nut	N/App	N/App	2.2 – 2.8	16 – 20
Cylinder 6 mm bolt	0.8 – 1.2	6 – 9	1.0 – 1.4	7 – 10
Cylinder 8 mm nut	2.2 – 2.8	16 – 20	2.2 – 2.8	16 – 20
Cam sprocket bolt	1.7 – 2.3	12 – 17	1.7 – 2.3	12 – 17
Valve adjuster locknut	0.8 – 1.2	6 – 9	1.5 – 1.8	11 – 13
Alternator stator bolt	0.8 – 1.2	6 – 9	N/App	N/App
Alternator stator base	0.9 – 1.3	7 – 9	N/App	N/App
Valve adjuster inspection cover	0.8 – 1.2	6 – 9	1.0 – 1.4	7 – 10
Carburettor mounting clip	0.3 – 0.5	2 – 4	0.3 – 0.5	2 – 4
Decompressor cable locknut	N/App	N/App	0.5 – 0.7	4 – 5
Selector drum stopper plate	1.0 – 1.4	7 – 10	N/App	N/App
Selector cam bolt	1.0 – 1.4	7 – 10	N/App	N/App
Gearbox sprocket	0.8 – 1.2	6 – 9	N/App	N/App
Engine oil drain plug	2.0 – 3.0	14 – 22	3.0 – 4.0	22 – 29
Spark plug	1.5 – 2.0	11 – 14	1.5 – 2.0	11 – 14

Specifications relating to Chapter 1 – XR models

Engine

Compression ratio – XR250 R 10.0:1
Power output:
    XR250 R .... 24.0 bhp @ 8500 rpm
    XR500 R .... 35.0 bhp @ 6500 rpm
Maximum torque:
    XR250 R .... 2.10 kg/cm (15.2 ft lb) @ 6500 rpm
    XR500 R .... 4.30 kg/cm (13.1 ft lb) @ 4000 rpm

Camshaft and rockers

Cam lift – XR500 R:
    Inlet .... 36.431 mm (1.4342 in)
    Service limit .... 36.23 mm (1.426 in)
    Exhaust .... 36.466 mm (1.4357 in)
    Service limit .... 36.27 mm (1.428 in)

Balancer assembly

Type – XR250 R .... Single gear-driven front balancer shaft

Gearbox

Type – XR250 R .... Six-speed constant mesh
Ratios – XR250 R:
    1st 3.000:1 (42/14T)
    2nd 2.111:1 (38/18T)
    3rd 1.591:1 (35/22T)
    4th 1.280:1 (32/25T)
    5th 1.074:1 (29/27T)
    6th 0.867:1 (26/30T)

Gear pinion internal diameters – XR250 R:
Output shaft 1st, 3rd, 4th and input shaft 5th, 6th	25.020 – 25.041 mm (0.9850 – 0.9859 in)
Service limit	25.10 mm (0.988 in)

Gear pinion bush dimensions:
Internal diameter	20.020 – 20.041 mm (0.7866 – 0.7890 in)
Outside diameter	25.005 – 25.016 mm (0.9844 – 0.9849 in)
Gear to bushing clearance – XR250 R	0.004 – 0.036 mm (0.0002 – 0.0014 in)
Input shaft OD	24.959 – 24.980 mm (0.9826 – 0.9849 in)
Gear to shaft clearance – XR250 R	0.020 – 0.054 mm (0.0008 – 0.0021 in)

Output shaft OD:
At splined area	24.959 – 24.980 mm (0.9826 – 0.9835 in)
At plain area	19.987 – 20.000 mm (0.7869 – 0.7874 in)

Selector fork ID – XR250 R:
Centre	12.000 – 12.018 mm (0.4724 – 0.4732 in)
Service limit	12.05 mm (0.474 in)
Right and left	15.000 – 15.021 mm (0.5906 – 0.5914 in)
Service limit	15.05 mm (0.593 in)

Selector fork claw thickness:
XR250 R	4.93 – 5.00 mm (0.194 – 0.197 in)
Service limit	4.5 mm (0.18 in)

Selector shaft OD:
Centre	11.966 – 11.984 mm (0.4711 – 0.4718 in)
Service limit	11.91 mm (0.496 in)
Right and left	14.966 – 14.984 mm (0.5892 – 0.5899 in)
Service limit	14.91 mm (0.587 in)

Selector drum OD:
XR250 R	11.966 – 11.984 mm (0.4711 – 0.4718 in)
Service limit	11.91 mm (0.469 in)

Selector drum bore ID:
XR250 R	12.000 – 12.027 mm (0.4724 – 0.4735 in)
Service limit	12.10 mm (0.476 in)

Kickstart pinion ID:
XR250 R	22.000 – 22.021 mm (0.8661 – 0.8670 in)
Service limit	22.10 mm (0.870 in)

Kickstart shaft OD:
XR250 R	21.959 – 21.980 mm (0.8645 – 0.8654 in)
Service limit	21.91 mm (0.863 in)

Torque settings

Component	XR250 R kgf m	lbf ft	XR500 R kgf m	lbf ft
Kickstart stopper plate	N/App	N/App	1.8 – 2.5	13 – 18
Kickstart anchor pin	N/App	N/App	2.2 – 3.0	16 – 22
Selector drum retainer	N/App	N/App	0.9 – 1.3	7 – 9
Cylinder head cover bolt	1.0 – 1.4	7 – 10	1.0 – 1.4	7 – 10
Cylinder head bolt	3.5 – 4.0	25 – 29	N/App	N/App
Cylinder head nut	N/App	N/App	2.2 – 2.8	16 – 20
Cam sprocket bolt	1.7 – 2.3	12 – 17	1.7 – 2.3	12 – 17
Cam chain tensioner bolts	1.0 – 1.4	7 – 10	1.0 – 1.4	7 – 10
Cylinder bolts	1.0 – 1.4	7 – 10	1.0 – 1.4	7 – 10
Cylinder nuts	N/App	N/App	2.2 – 2.8	16 – 20
Crankcase cover bolt	0.8 – 1.2	6 – 9	N/App	N/App
Primary drive pinion nut	4.5 – 6.0	33 – 43	N/App	N/App
Automatic timing unit bolt	N/App	N/App	4.5 – 6.0	33 – 43
Clutch centre nut	4.5 – 6.0	33 – 43	4.5 – 6.0	33 – 43
Alternator stator bolt	0.8 – 1.2	6 – 9	N/App	N/App
Alternator rotor bolt	9.5 – 10.5	69 – 76	10.0 – 12.0	72 – 87
Balancer adjuster lock bolt	2.2 – 2.8	16 – 20	2.2 – 2.8	16 – 20

Specifications relating to Chapter 2

Fuel tank

Overall capacity – XR models	9.0 litre (2.4/1.9 US/Imp gal)
Reserve capacity – XR models	3.0 litre (0.8/0.6 US/Imp gal)

Carburettor

	XL250 R-C (UK) XL250 R (US)	XL500 R-C (UK)	XL500 R (US)
Make	Keihin	Keihin	Keihin
Venturi size	28 mm (1.10 in)	32 mm (1.26 in)	32 mm (1.26 in)
ID number	PD74A	PD78A	PD77A
Float level	14.0 mm (0.55 in)	18.0 mm (0.71 in)	18.0 mm (0.71 in)
Pilot screw (turns out)	$1\frac{3}{4}$	$2\frac{1}{4}$	$2\frac{1}{4}$

Carburettor (continued)

	XL250 R-C (UK) XL250 R (US)	XL500 R-C (UK)	XL500 R (US)
Idle speed	1200 ± 100 rpm	1200 ± 100 rpm	1200 ± 100 rpm
Main jet	110	130	128 or 135
Optional high altitude main jet	N/App	N/App	130
Throttle valve diameter	28 mm (1.10 in)	34 mm (1.34 in)	34 mm (1.34 in)
Pilot jet	38	55	52
Throttle free play	2 – 6 mm ($\frac{1}{8}$ – $\frac{1}{4}$ in)	2 – 6 mm ($\frac{1}{8}$ – $\frac{1}{4}$ in)	2 – 6 mm ($\frac{1}{8}$ – $\frac{1}{4}$ in)
Fast idle speed	2000 – 2500 rpm	N/Av	N/Av
Accelerator pump delivery per stroke	0.1 – 0.25 cc	0.1 – 0.25 cc	0.1 – 0.25 cc
Air cut-off valve operating pressure	350 – 430 mm Hg (13.8 – 16.9 in Hg)	350 – 430 mm Hg (13.8 – 16.9 in Hg)	350 – 430 mm Hg (13.8 – 16.9 in Hg)

Carburettor

	XR250 R	XR500 R
Make	Keihin	Keihin
ID number	PD71A	PD11B
Jet needle position	3rd groove	3rd groove
Float level	12.5 mm (0.49 in)	14.5 mm (0.57 in)
Pilot screw (turns out)	$2\frac{1}{2}$	$2\frac{1}{4}$
Idle speed	1300 ± 100 rpm	1200 ± 100 rpm
Main jet	130	152
Optional high altitude main jet	See text	See text
Throttle valve diameter	30.5 mm (1.10 in)	N/Av
Pilot jet	40	55
Throttle free play	2 – 6 mm ($\frac{1}{8}$ – $\frac{1}{4}$ in)	2 – 6 mm ($\frac{1}{8}$ – $\frac{1}{4}$ in)
Air cutoff valve operating pressure	360 mm Hg (14.2 in Hg)	390 mm Hg (15.3 in Hg)

Specifications relating to Chapter 3

Ignition system – UK XL250/500 R-C, US XL250/500 R

Type .... Capacitor discharge (CDI), electronic ignition advance

Spark plug – UK XL250/500 R-C, US XL250/500 R

	NGK	ND
Make		
Type:		
Standard	DR8ES-L	X24ESR-U
Cold climate (below 5°C, 41°F)	DR7ES	X22ESR-U
Continuous high-speed riding	DR8ES	X27ESR-U
Electrode gap	0.6 – 0.7 mm (0.024 – 0.028 in)	

Ignition timing – UK and US XL models

	XL250 R, R-C	XL500 R, R-C
Initial	12° BTDC @ 1200 rpm	10° BTDC @ 1200 rpm
At full advance	37° BTDC @ 4000 rpm	35° BTDC @ 3500 rpm

Ignition system – XR models

Type .... Capacitor discharge (CDI), mechanical ignition advance

Spark plug – XR models

	US spec models		Canadian spec models	
	NGK	ND	NGK	ND
Make				
Type – XR250 R:				
Standard	D8EA	X24ES-U	DR8ES-L	X24ESR-U
Cold climate below 5°C (41°F)	D7ES	X22ES-U	DR7ES	X22ESR-U
Continuous high-speed riding	D9EA	X27ES-U	DR8ES	X27ESR-U
Type – XR500 R:				
Standard	D8EA	X24ES-U	DR8ES-L	X24ESR-U
Electrode gap	0.6 – 0.7 mm (0.024 – 0.028 in)			

Ignition timing – XR models

Initial	12° BTDC @ 2250 ± 250 rpm
At full advance:	
XR250 R	37° ± 2° BTDC @ 2900 rpm
XR500 R	36° ± 2° BTDC @ 3500 rpm

Specifications relating to Chapter 4

Frame

	XL250 S-B UK model
Type	Welded tubular steel

Front forks

Type	
Travel	
Oil capacity per leg	
Oil grade	
Fork spring free length	
Service limit	
Lower leg ID	
Fork stanchion bend limit	

XL250 S-B UK model
Oil damped telescopic
204.0 mm (8.0 in)
190 cc (6.7/6.4 Imp/US fl oz)
ATF (automatic transmission fluid)
562.4 mm (22.1 in)
551.0 mm (21.7 in)
44.2 mm (1.74 in)
0.2 mm (0.008 in)

Rear suspension

Type	
Travel	
Suspension unit spring free length	
Service limit	
Swinging arm bush clearance	

Swinging arm with twin oil damped suspension units
178 mm (7.0 in)
326.9 mm (12.87 in)
320.5 mm (12.61 in)
0.2 – 0.3 mm (0.008 – 0.012 in)

Frame

Type

XL250/500 R-C UK, XL250/500 R US
Welded tubular steel

Front forks

Type	
Travel	
Oil capacity per leg:	
XL250 R/R-C	
XL500 R/R-C	
Oil level:	
XL250 R/R-C	
XL500 R/R-C	
Oil grade	
Fork air pressure	
Fork spring free length:	
XL250 R/R-C	
XL500 R/R-C	
Service limit:	
XL250 R/R-C	
XL500 R/R-C	
Fork stanchion bend limit	

Oil damped telescopic, air assisted
215.0 mm (8.5 in)

300 cc (10.6/10.1 Imp/US fl oz)
378.5 cc (13.3/12.8 Imp/US fl oz)

173 mm (6.81 in)
163 mm (6.42 in)
ATF (automatic transmission fluid)
0 – 0.2 kg/cm^2 (0 – 2.8 psi)

579.9 mm (22.83 in)
580.4 mm (22.85 in)

568.3 mm (22.37 in)
568.8 mm (22.39 in)
0.2 mm (0.008 in)

Rear suspension

Type	
Travel	
Suspension unit spring free length:	
XL250 R/R-C	
XL500 R/R-C	
Service limit:	
XL250 R/R-C	
XL500 R/R-C	
Damper unit compression pressure (without spring)	
Service limit	
Swinging arm pivot centre collar OD	
Service limit	

Rising rate (Honda Pro-link) with single gas-filled suspension unit
190 mm (7.5 in)

246.5 mm (9.70 in)
251.0 mm (9.88 in)

244.0 mm (9.56 in)
248.5 mm (9.78 in)
28 – 38 kg (62 – 84 lb)
28 kg (62 lb)
19.987 – 20.000 mm (0.7869 – 0.7874 in)
19.915 mm (0.07840 in)

Frame

Type

XR250 R	**XR500 R**
Welded tubular steel	

Front forks

	XR250 R	**XR500 R**
Type		
Travel		
Oil capacity per leg:		
Standard 1981		
Standard 1982		
Oil level:		
Standard – 1981 model		
Standard – 1982 model		
Maximum – 1981 model		
Maximum – 1982 model		
Minimum – 1981 model		
Minimum – 1982 model		

	XR250 R	**XR500 R**
Type	Oil damped telescopic, air assisted	
Travel	255.0 mm (10.0 in)	255.0 mm (10.0 in)
Standard 1981	368 cc (13.0/12.4 Imp/US fl oz)	345 – 350 cc (12.1 – 12.3/ 11.6 – 11.8 Imp/US fl oz)
Standard 1982	395 cc (13.9/13.4 Imp/US fl oz)	395 cc (13.9/13.4 Imp/US fl oz)
Standard – 1981 model	152 mm (6.00 in)	181 mm (7.10 in)
Standard – 1982 model	156 mm (6.14 in)	156 mm (6.14 in)
Maximum – 1981 model	132 mm (5.20 in)	161 mm (6.30 in)
Maximum – 1982 model	146 mm (5.75 in)	146 mm (5.75 in)
Minimum – 1981 model	172 mm (6.77 in)	201 mm (8.00 in)
Minimum – 1982 model	186 mm (7.32 in)	186 mm (7.32 in)

Front forks (continued)

	XR250 R	XR500 R
Oil grade	ATF (automatic transmission fluid)	ATF
Fork air pressure:		
Standard – 1981 model	0.4 kg/cm² (5.7 psi)	0.3 kg/cm² (4.3 psi)
Standard – 1982 model	0 kg/cm² (0 psi)	0 kg/cm² (0 psi)
Maximum	1.0 kg/cm² (14.2 psi)	1.0 kg/cm² (14.2 psi)
Fork spring free length	562.4 mm (22.10 in)	617.5 mm (23.31 in)
Service limit	551.0 mm (21.70 in)	605.1 mm (24.00 in)
Fork stanchion bend limit	0.2 mm (0.008 in)	0.2 mm (0.008 in)

Rear suspension

	XR250 and 500 R
Type:	
XR250 R	Rising rate (Honda Pro-link) with single gas-filled suspension unit
XR500 R	Rising rate (Honda Pro-link) with single remote reservoir suspension unit
Travel	255 mm (10.0 in)
Suspension unit spring free length	222.6 mm (8.76 in)
Service limit	219.0 mm (8.62 in)
Damper unit compression pressure (without spring)	28 – 38 kg (62 – 84 lb)
Service limit	23 kg (51 lb)

Torque settings

Component	UK XL250/500 R-C, US XL250/500 R kgf m	lbf ft
Engine mounting bolts:		
8 mm	3.0 – 3.7	22 – 27
10 mm – 250 cc	5.0 – 6.5	36 – 47
10 mm – 500 cc	5.5 – 6.5	40 – 47
12 mm	9.0 – 10.0	65 – 72
Steering stem nut	8.0 – 12.0	58 – 87
Steering stem adjuster	0.1 – 0.2	0.7 – 1.4
Handlebar upper holder bolt	1.8 – 3.0	13 – 22
Steering stem pinch bolt	4.0 – 5.0	29 – 36
Fork yoke pinch bolt:		
Upper	1.8 – 2.3	13 – 17
Lower	3.0 – 3.5	22 – 25
Front wheel spindle	5.0 – 8.0	36 – 58
Front wheel spindle holder nut	1.0 – 1.4	7 – 10
Rear wheel spindle	8.0 – 11.0	58 – 80
Swinging arm pivot	7.0 – 10.0	51 – 72
Suspension unit mounting:		
Upper	6.0 – 7.5	43 – 54
Lower	3.8 – 4.8	28 – 35
Suspension linkage pivots:		
Swinging arm to suspension unit arm	9.0 – 12.0	65 – 87
Suspension unit arm to link	6.0 – 7.5	43 – 54
Link to frame	6.0 – 7.5	43 – 54
Footrest bracket	7.0 – 10.0	51 – 72
Kickstart lever	2.0 – 3.5	15 – 25
Gearchange pedal	0.8 – 1.2	6 – 9
Exhaust retainer	1.5 – 2.5	11 – 18
Silencer mounting	2.0 – 3.0	15 – 22
Side stand pivot	3.5 – 4.5	25 – 33

Torque settings

Component	XR250 R kgf m	lbf ft	XR500 R kgf m	lbf ft
Fuel tank	N/Av	N/Av	1.5 – 2.4	11 – 17
Exhaust pipe flange	N/Av	N/Av	0.8 – 1.2	6 – 9
Exhaust heat shield	N/Av	N/Av	0.8 – 1.2	6 – 9
Handlebar holder	1.8 – 3.0	13 – 22	1.8 – 3.0	13 – 22
Steering stem	7.0 – 10.0	51 – 72	8.8 – 12.0	58 – 87
Fork yoke pinch bolt:				
Upper	1.8 – 3.0	13 – 22	1.8 – 2.5	13 – 18
Lower	1.8 – 3.0	13 – 22	1.8 – 3.0	13 – 22
Front wheel spindle	5.0 – 8.0	36 – 58	5.0 – 8.0	36 – 58
Front wheel spindle holder	1.0 – 1.2	7 – 9	1.0 – 1.4	7 – 10
Damper rod Allen bolt	1.5 – 2.5	11 – 18	0.8 – 1.2	6 – 9
Rear suspension unit mountings:				
Upper	6.0 – 7.5	43 – 54	7.0 – 10.0	51 – 72
Lower	3.8 – 4.8	27 – 35	3.8 – 4.8	27 – 35

Torque settings (continued)

Component	XR250 R kgf m	lbf ft	XR500 R kgf m	lbf ft
Suspension linkage:				
Tension arm	9.0 – 12.0	65 – 87	9.0 – 12.0	65 – 87
Tension rod	6.0 – 7.5	43 – 54	6.0 – 7.5	43 – 54
Swinging arm pivot	7.0 – 10.0	51 – 72	7.0 – 10.0	51 – 72
Rear wheel spindle	7.0 – 11.0	51 – 80	8.0 – 11.0	58 – 80
Engine mountings:				
8 mm	2.5 – 3.5	14 – 25	4.5 – 6.0	33 – 43
10 mm upper	4.5 – 6.0	33 – 43	N/App	N/App
10 mm front	3.0 – 5.0	22 – 37	N/App	N/App
10 mm rear	8.0 – 9.5	58 – 69	N/App	N/App
10 mm – XR500 R	N/App	N/App	3.0 – 5.0	22 – 36
12 mm – XR500 R	N/App	N/App	10.0 – 13.0	72 – 94
Gearchange pedal	0.8 – 1.2	6 – 9	0.8 – 1.2	6 – 9
Kickstart lever	2.0 – 3.5	14 – 25	2.0 – 3.5	14 – 25
Side stand	3.5 – 4.5	22 – 33	3.5 – 4.5	22 – 33

Specifications relating to Chapter 5

Brakes

XL250 S-B (UK) XL250 S 1981 (US)

Swept area – rear .... 122.5 cm² (19.0 in²)

Tyres

XL250/500 R-C UK, XL250/500 R US

Size:
Front 3.00-21-4PR
Rear 4.60-17-4PR
Pressures:
Front .... 21 psi (1.5 kg cm²)
Rear (UK):
XL250 R-C .... 21 – 24 psi (1.5 – 1.75 kg/cm²)
XL500 R-C .... 21 psi (1.5 kg/cm²)
Rear (US) .... 21 psi (1.5 kg cm²)

Brakes

Type:
Front:
XL250R/R-C .... Single leading shoe (sls) drum brake
XL500/R-C .... Twin leading shoe (tls) drum brake
Rear .... Single leading shoe (sls) drum brake
Swept area:
Front .... 102 cm² (15.9 in²)
Rear:
XL250 R/R-C .... 104 cm² (16.1 in²)
XL500 R/R-C .... 122 cm² (19.0 in²)

Tyres

XR250/500 R

Size:
Front 3.00-21-6PR
Rear 5.10-17-6PR
Pressures:
Front .... 14 psi (1.0 kg cm²)
Rear .... 11 psi (0.8 kg cm²)

Brakes

Type:
Front .... Twin leading shoe (tls) drum brake
Rear .... Single leading shoe (sls) drum brake
Swept area:
Front .... 102 cm² (15.9 in²)
Rear:
XR250 R .... 104 cm² (16.1 in²)
XR500 R .... 122 cm² (19.0 in²)

XL250/500 R-C (UK)
XL250/500 R (US)

Torque settings

Component	XR500 R kgf m	lbf ft	XR250 R kgf m	lbf ft
Front wheel spindle	5.0 – 8.0	36 – 58	5.0 – 8.0	36 – 58
Front wheel spindle holder	1.0 – 1.4	7 – 10	1.0 – 1.2	7 – 9
Rear wheel spindle	8.0 – 11.0	58 – 80	7.0 – 11.0	51 – 80
Rear wheel sprocket	2.8 – 3.4	20 – 25	2.7 – 3.3	20 – 24

Specifications relating to Chapter 6

Generator
Type ....
Rating ....
Charging commences ....
Charging output (unregulated):
    Minimum ....
    Maximum ....
Lighting output (unregulated):
    Minimum (250) ....
    Minimum (500) ....
    Maximum (250) ....
    Maximum (500) ....

UK XL250/500 R-C	US XL250/500 R
Alternator	Alternator
12V 0.196kW @ 5000 rpm	N/Av
1200 rpm or less	1000 rpm or less
16.8V, 2.7A @ 2500 rpm	16.8V, 2.7A @ 2500 rpm
18.4V, 5.5A @ 8000 rpm	18.4V, 5.5A @ 8000 rpm
14.5V @ 2500 rpm	14.5V @ 2500 rpm
13.0V @ 2500 rpm	14.5V @ 2500 rpm
16.0V @ 8000 rpm	23.0V @ 8000 rpm
23.0V @ 8000 rpm	23.0V @ 8000 rpm

Battery
Capacity ....

UK XL250/500 R-C, US XL250/500 R

12V 3Ah

Fuse
Rating ....

10A

Regulator/rectifier
Type ....
Battery charge rate ....

Transistorised
0.3A max

Bulb wattages

	UK XL250/500 R-C	US XL250/500 R
Headlamp	35/35W	35/36.5W
Stop/tail	5/21W	8/27W
Turn signal	21W	23W
Speedometer	1.7W	1.7W
Tachometer	3.4W	N/Av
Neutral indicator	3.4W	3.4W
Turn signal warning	3.4W	3.4W
Main beam warning	1.7W	1.7W
Parking lamp	4.0W	N/App

Generator
Type ....
Rating ....

XR250/500 R
Alternator
6V 50W @ 5000 rpm

Bulb wattages
Headlamp ....
Tail ....

6V 25W
6V 3W

---

1 Introduction

This update chapter covers the Honda XL and XR 250 and 500 models produced between 1981 and 1984, other than those described in the original text. It should be noted that model references applied to the early machines and quoted in the main part of this manual may have changed retrospectively. For example, the orginal UK XL250 was known as the XL250 S. It has subsequently been referred to as the XL250 S-Z to distinguish it from the later XL250 S-A. To avoid confusion, a summary of the various models, together with introduction and discontinuation dates is shown below.

In many cases, models continued to be sold well after they were officially discontinued. It follows that the only certain way of making sure that correct replacement parts are supplied is to specify the engine and frame numbers when ordering. This point is especially true where model runs are long and numerous modifications have been made to the range.

UK models

Model	Introduced	Discontinued	Coverage
XL250 S-Z	May 1978	April 1980	Chapters 1 to 6
XL250 S-A	Jan 1980	May 1981	Chapters 1 to 6
XL250 S-B	May 1981	Sept 1982	Chapter 7
XL250 R-C	Mar 1982	Oct 1984	Chapter 7
XR250 Z	Mar 1979	N/Av	Chapters 1 to 6
XR250 A	See note 1 below		Chapter 7
XL500 S-Z	Mar 1979	End 1981	Chapters 1 to 6
XL500 S-A	End 1981	Feb 1982	Chapters 1 to 6
XL500 R-C	Feb 1982	Oct 1984	Chapter 7
XR500 Z	Mar 1979	N/Av	Chapters 1 to 6
XR500 A	See note 2 below		Chapter 7

US models

Model	Introduced	Discontinued	Coverage
XL250 S	April 1978	1980	Chapters 1 to 6
XL250 S	1981	1981	Chapter 7
XL250 R	Feb 1982	1983	Chapter 7
XR250	Sept 1978	1980	Chapters 1 to 6
XR250 R	Sept 1981	1982	Chapter 7
XL500 S	Feb 1979	1980	Chapters 1 to 6
XL500 S	1981	1981	Chapter 7
XL500 R	July 1981	1982	Chapter 7
XR500	Jan 1979	1980	Chapters 1 to 6
XR500 R	Aug 1981	1982	Chapter 7

Note 1: The XR250 A model sold in the UK was imported from Canada by Honda UK, and thus conforms to Canadian specifications. It is covered as far as is practicable by this update.

Note 2: The XR500 models were imported from Canada by specialist dealers in the UK, and is thus a Canadian specification model. It is suggested than when attempting to order spare parts this should be done via the dealer who supplied the machine; there may be difficulty in attempting to obtain parts from normal Honda stockists.

Notes: This manual **does not** cover any of the RFVC engine models. These were produced from 1983 onwards (though most models were introduced during 1984) so there is some overlap between these and the earlier machines covered here.

2 Balancer shaft modifications: general

A revised balancer arrangement is fitted to all models equipped with a six-speed gearbox; namely the UK XL250 R-C and the US XL250 R and XR250 R models. The chain-driven twin balancer arrangement used on the other models of the range is replaced by a single gear-driven balancer at the front of the crankcase, the gear drive being taken from the left-hand end of the crankshaft. The space occupied by the rear balancer weight is then utilised to allow the fitting of a sixth gear.

3 Balancer shaft: removal and refitting – 6-speed models

1 After separating the crankcase halves as described in Chapter 1, release the circlip which retains the balancer weight to the right-hand end of the balancer shaft assembly. The weight can now be slid off its splines and the shaft displaced from the holder. Slide the holder out of the crankcase bore. If

necessary, the holder flange can be removed after releasing the circlip which retains it.

2 The needle roller bearings are located on the balancer shaft by circlips and may be removed once the clips have been slid off. When refitting the bearings and clips, and also the holder flange, note that the circlips have a rounded edge and a sharp edge; the sharp edge must face outwards.

3 Fit the holder into the crankcase bore and slide the balancer shaft into place. Fit the balancer weight to the shaft, ensuring that the punch marks on the shaft end and the weight align, then fit the circlip which retains it.

4 When joining the crankcase halves, check that the driven gear is positioned so that the alignment mark on its outer face coincides with the punch mark on the driving gear. Before proceeding further, turn the crankshaft to TDC and check that the flat edge of the balancer weight aligns with the crankcase alignment mark. If this is not the case, go through the assembly sequence again, checking all alignment marks.

4 Balancer shaft: backlash adjustment – 6-speed models

The eccentric balancer shaft holder allows the backlash (free play) between the drive and driven gears to be set and adjusted as required. Adjustment should not be required frequently, but it should be checked during engine reassembly and whenever the balancer system appears unusually noisy in operation. Start by slackening the balancer holder flange lock bolt. Turn the holder flange until the gear teeth contact lightly, then set the holder flange 1.5 – 2.0 graduations to the left in relation to the crankcase mark. Once the backlash has been set correctly, tighten the lock bolt to 2.0 – 2.6 kgf m (14 – 19 lbf ft).

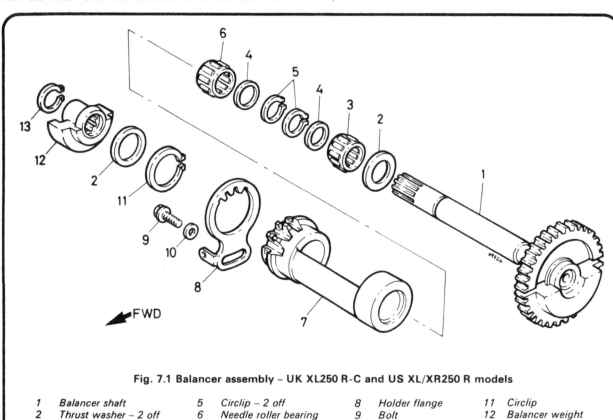

Fig. 7.1 Balancer assembly – UK XL250 R-C and US XL/XR250 R models

1 Balancer shaft	5 Circlip – 2 off	8 Holder flange	11 Circlip
2 Thrust washer – 2 off	6 Needle roller bearing	9 Bolt	12 Balancer weight
3 Needle roller bearing	7 Holder	10 Special.washer	13 Circlip
4 Washer – 2 off			

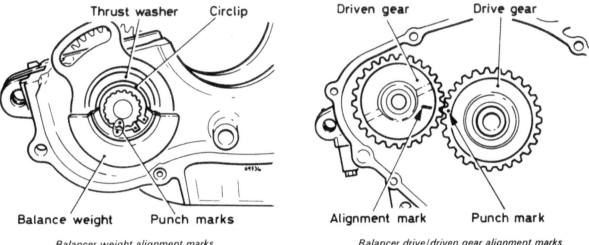

Balancer weight alignment marks

Balancer drive/driven gear alignment marks

Fig. 7.2 Balancer weight and drive gear installation – UK XL250 R-C and US XL/XR250 R models

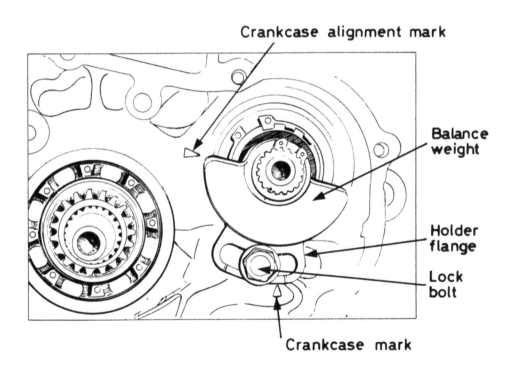

Fig. 7.3 Balancer gear backlash adjustment

5   Gearbox modifications: general – UK XL250 R-C and US XL250 R/XR250 R models

1   The above-mentioned models are fitted with a 6-speed gearbox in place of the 5-speed assembly used in the rest of the range. The additional pair of gears necessitated utilising space in the crankcase normally occupied by the rear balancer weight, and thus the balancer arrangement was changed as described in the preceding sections.

2   The change to the new gear clusters has not materially affected the approach to dismantling and reassembly, and the general remarks in Chapter 1 can be applied. Note, however, that the order of the gear pinions on the shafts is changed, and thus the assembly sequence outlined in section 29 of Chapter 1 cannot be applied literally. An exploded view of the two gear clusters together with the various washers and circlips accompanies this Section; use this as a guide during dismantling and reassembly.

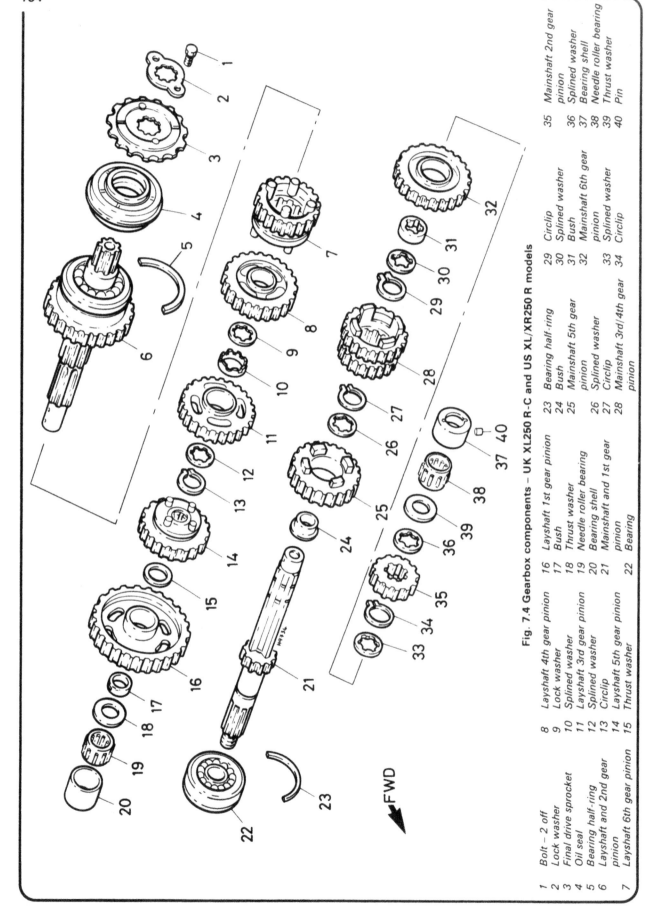

Fig. 7.4 Gearbox components – UK XL250 R-C and US XL/XR250 R models

1	Bolt – 2 off	8	Layshaft 1st gear pinion
2	Lock washer	9	Bush
3	Final drive sprocket	10	Thrust washer
4	Oil seal	11	Layshaft 3rd gear pinion
5	Bearing half-ring	12	Splined washer
6	Layshaft and 2nd gear	13	Circlip
	pinion	14	Layshaft 5th gear pinion
7	Layshaft 6th gear pinion	15	Thrust washer

16	Layshaft 4th gear pinion	23	Bearing half-ring
17	Lock washer	24	Bush
18	Splined washer	25	Mainshaft 5th gear
19	Needle roller bearing		pinion
20	Bearing shell	26	Splined washer
21	Mainshaft and 1st gear	27	Circlip
	pinion	28	Mainshaft 3rd/4th gear
22	Bearing		pinion

29	Circlip	35	Mainshaft 2nd gear
30	Splined washer		pinion
31	Bush	36	Splined washer
32	Mainshaft 6th gear	37	Bearing shell
	pinion	38	Needle roller bearing
33	Splined washer	39	Thrust washer
34	Circlip	40	Pin

6   Automatic cam chain tensioner: general – all UK R-C and US R-models

1   A revised cam chain tensioner arrangement is fitted to all models from 1982 onwards. This applies to the UK XL250 R-C and XL500 R-C and to the US XL/XR 250 and 500 R models. In addition, any late model XR machine imported into the UK from Canada will feature this arrangement.

2   In the case of the earlier models, the cam chain tension was set after assembly, when the engine was first run. The tensioner, having moved to the correct position, was then locked. Over a period of time the cam chain would wear and stretch slightly, necessitating readjustment.

3   On later models the tensioner was modified to operate automatically, taking up additional chain free play as it developed. A system of wedges locks the tensioner in its new position to prevent it from backing off again. The arrangement is shown in section in the accompanying illustration.

4   Although no maintenance is required, there are occasions when it is necessary to manually back off the tensioner, either to facilitate camshaft removal or simply to check the operation of the mechanism; it is not unknown for the mechanism to jam in one position. This can be accomplished with the cylinder head cover removed.

5   Grasp the top of the tensioner wedge with a pair of pliers and pull it upwards against spring pressure. At the same time push down on the stopper wedge to free off the tensioner. To lock the tensioner in its off position, insert a pin into the hole in the tensioner wedge to hold it up. To reset the tensioner, remove the pin from the tensioner wedge and push down on the stopper wedge to allow the mechanism to assume normal operation. The mechanism and the locking method are shown in the accompanying line drawing.

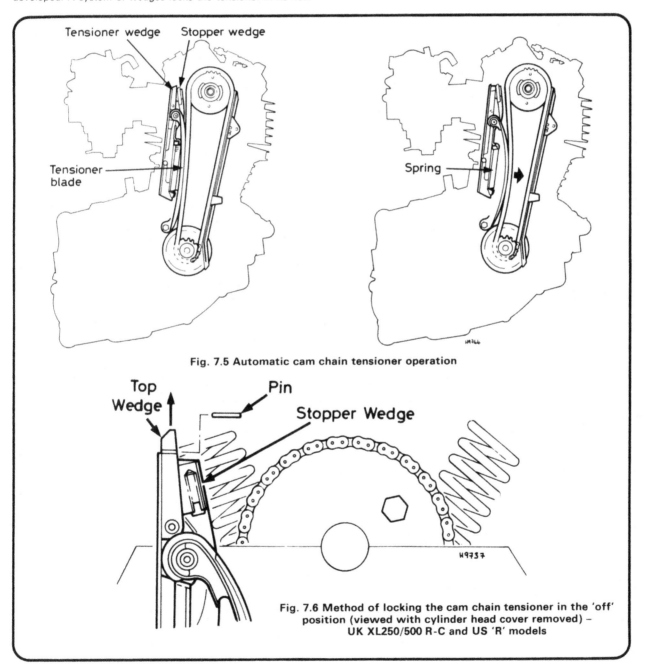

Fig. 7.5 Automatic cam chain tensioner operation

Fig. 7.6 Method of locking the cam chain tensioner in the 'off' position (viewed with cylinder head cover removed) – UK XL250/500 R-C and US 'R' models

7   Clutch modifications: general – UK XL250 S-B, R-C and US XL250 S ('81), XL250 R models

1   The clutch arrangement used on the above models differs slightly from that fitted to the earlier machines, and this can be seen by comparing the accompanying illustration with that shown in Fig. 1.4. The design of the clutch outer drum is modified, the slots being deeper to accommodate the extra clutch plates.

2   The new type clutch uses five friction and four plain plates. A large thrust washer is fitted between the clutch pressure plate and the outer drum. The clutch centre is fitted with an anti-judder spring and seat similar to that used on the earlier XL500 models. The clutch release mechanism remains unchanged apart from the pushrod; this no longer has a spigot extending through the release bearing, but presents a flat mushroom head to the bearing inner race.

8   Clutch modifications: general – UK XL500 R-C and US XL500 R models

The clutch used on the later XL500 models is largely unchanged from earlier versions, and in fact it is this clutch which forms the basis for the arrangement used on all of the later models. The only significant change is to the clutch release pushrod which, like that of the 250 models, has lost the extended spigot used previously. For details refer to the illustration of the XL250 R/R-C clutch.

9   Clutch modifications: general – XR250/500 R models

In the case of the later XR models, the 250cc version uses a clutch derived from the original XL500, and this is identical to the XL250 R/R-C clutch described above. In the case of the XR500 R the clutch is unchanged from earlier versions and reference should be made to Chapter 1 and Fig. 1.5.

10   Ignition pickup modifications: general – US/UK XL250/500 R and R-C models

Whilst the XR models retain the centrifugal ignition advance arrangement of their predecessors, this system was changed in the case of the XL250 and XL500 R/R-C machines. The automatic timing unit (see photograph 38.4a in Chapter 1 and photograph 8.1b in Chapter 3) is omitted, ignition advance being carried out electronically. The new type rotor is retained in the same way to the end of the crankshaft, this being otherwise unchanged.

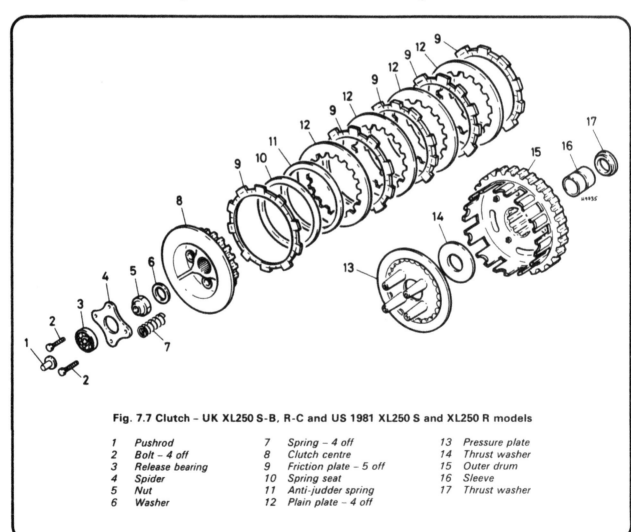

Fig. 7.7 Clutch – UK XL250 S-B, R-C and US 1981 XL250 S and XL250 R models

1	Pushrod	7	Spring – 4 off	13	Pressure plate
2	Bolt – 4 off	8	Clutch centre	14	Thrust washer
3	Release bearing	9	Friction plate – 5 off	15	Outer drum
4	Spider	10	Spring seat	16	Sleeve
5	Nut	11	Anti-judder spring	17	Thrust washer
6	Washer	12	Plain plate – 4 off		

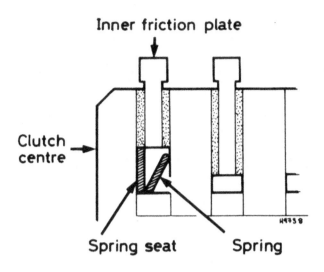

Fig. 7.8 Correct installation of clutch anti-judder spring

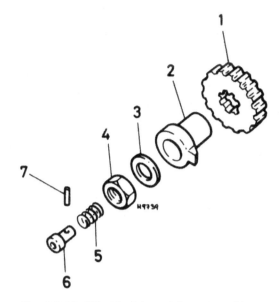

Fig. 7.9 Modified ignition pickup assembly –
UK XL250/500 R-C and US XL250/500 R models

1	Primary drive pinion	5	Spring
2	Rotor	6	Oil feed quill
3	Washer	7	Pin
4	Nut		

11 Carburettor modifications: general

The machines covered in this update Chapter all feature some changes to the carburettor, these mostly relating to changes in jet size and adjustment settings. Information relating to this will be found in the Specifications at the beginning of the Chapter, and this should be referred to when working on the instruments concerned. In some instances, more significant alterations have been made, and these are discussed in more detail in the sections which follow.

12 Carburettor: settings and adjustments

UK XL250 R-C model
1   The UK XL250 R-C features the accelerator pump assembly which had been used on XL250 models in other countries for some time. Unfortunately, full details of this assembly were not available at the time of writing, so for the purposes of this section it has been assumed that the arrangement is as specified for the US models.
2   An exploded view of the carburettor is shown in Chapter 2 in Fig. 2.2, and this can be used as reference when dealing with the carburettor on the R-C model. Note, however, that the components marked 'U.S. type only' are now applicable to the UK machine. Dismantling and reassembly of the carburettor, including the accelerator pump components is discussed in Section 6 of Chapter 2. The only significant change is to the slow, or pilot jet. On all previous machines this was a press-fit in the carburettor body and thus could not be removed. In the case of the R-C model the jet is screwed into place and has a slotted head to permit removal for cleaning purposes.

US XL250 R 1982 model
3   The 1982 XL250 R model carburettor differs from earlier machines in detail only (see specifications). The exception to this is optional main jet settings for high altitude operation. Honda advise that if the machine is to be used continuously at an altitude of 6500 feet (2000 metres) or more, the standard 110 main jet should be removed and a 105 main jet fitted in its place. Failure to do so will affect the running of the machine,

possibly causing overheating, and will also increase exhaust emissions. After the new jet has been fitted, check the idle speed and reset, if required, to 1200 ± 100 rpm. Where local leglislation requires it, it may be necessary to fit a Vehicle Emission Control Update label, in which case a Honda dealer should be consulted before any change is made.

UK XL500 R-C model
4   Like the 250cc model, the XL500 R-C saw the introduction of an accelerator pump and a removable pilot (slow) jet to the UK version. The remarks in paragraphs 1 and 2 above can thus be applied to the 500cc model. In addition there is information relating to the adjustment of the accelerator pump, pilot screw and fast idle speed, which should be carried out as follows.
5   To check the accelerator pump adjustment, slacken off the throttle stop screw so that the throttle valve closes fully. It is a good idea to count the number of turns and part turns so that the original setting can be approximated. Using feeler gauges, measure the clearance between the end of the pump rod and tang on the arm with which it engages. There should be a gap of 0.0 – 0.5 mm (0.000 – 0.002 in).
6   If necessary, adjust this setting by carefully bending the tang with pliers. Note that the engine idle speed setting will have been lost when the throttle stop screw was slackened, and this should be reset when the engine is at normal operating temperature to 1200 ± 100 rpm. In the absence of a tachometer, set the idle speed to the slowest reliable setting.
7   Honda recommend the following procedure for setting the pilot screw after overhaul. They point out that the screw is set correctly during assembly and should not normally be disturbed, but should the setting have been lost, proceed as follows. First set the screw to the initial position of 2¼ turns out. When doing this take care not to screw the pilot screw hard into its seating or damage will be caused.
8   Ride the machine normally for 10 – 15 minutes to allow the engine to reach full working temperature, then stop it and connect a test tachometer. Start the engine and set the idle

speed to the recommended 1200 $\pm$ 100 rpm. With the engine idling, turn the screw clockwise in $\frac{1}{4}$ turn increments until the engine stalls. Back off the screw by 2 complete turns. This will be the correct screw setting. Re-start the engine and if necessary reset the idle speed using the throttle stop screw.
9 The fast idle speed is set with the engine warm and the choke knob at the intermediate detent position. If the resulting engine speed is outside the range 2000 – 2500 rpm, adjust it using the spring-loaded adjuster bolt, having first slackened off the locknut.

US XL500 R 1982 model
10 The 1982 XL500 R model carburettor differs from earlier machines in detail only (see specifications). The exception to this is optional main jet settings for high altitude operation. Honda advise that if the machine is to be used continuously at an altitude of 6500 feet (2000 metres) or more, the standard 135 main jet should be removed and a 130 main jet fitted in its place. Failure to do so will affect the running of the machine, possibly causing overheating, and will also increase exhaust emissions. After the new jet has been fitted, check the idle speed and reset if required to 1200 $\pm$ 100 rpm. Where local legislation requires it, it may be necessary to fit a Vehicle Emission Control Update label, in which case a Honda dealer should be consulted before any change is made. **Note:** *Machines with frame number through JH2PD0200CM006317 were manufactured with a 128 main jet.*

XR250 R model
11 The information relating to the XR250 R applies primarily to the US model, but can also be applied to any Canadian specification machines imported into the UK. In general, changes are confined to jet sizes and settings and these are covered in the specifications at the beginning of this Chapter.
12 To obtain optimum performance and reliability under competition conditions, the carburation must be corrected for local temperature and altitude variations as follows. Start by determining the ambient temperature and altitude, and then use the accompanying chart to establish the required correction factor (C). If C is 0.95 or below it is necessary to make alterations to the jet needle position and pilot screw setting; lower the jet needle by one groove and turn the pilot screw out by a $\frac{1}{2}$ turn from the standard setting of $2\frac{1}{2}$ turns out. If C is above 0.95 these alterations are not required.
13 To determine the required jet size, multiply the standard jet size (130) by the correction factor C. The resulting figure gives an optimum jet size; choose the nearest actual jet from the optional range (see below).
14 As an example of the above process, at a temperature of 30°C (86°F) and at an altitude of 3000m (9840ft) a correction factor (C) of 0.92 is indicated. The adjustment would be as follows:

Main jet: 130 (std) x 0.92 = 119.6 (120 being the nearest actual jet size)
Jet needle position: 3rd groove −1 = 2nd groove
Pilot screw settings: $2\frac{1}{2} - \frac{1}{2} = 2$ turns out

15 The following optional main jets can be obtained from Honda dealers. In the case of UK models these are unlikely to be available from normal Honda dealers, in which case consult the dealer who first supplied the machine, or a specialist in Honda competition machines. Optional main jets:

Size	Part Number
115	99101-357-115
118	99101-357-118
120	99101-357-120
122	99101-357-122
125	99101-357-125
128	99101-357-128
132	99101-357-132
135	99101-357-135

XR500 R model
16 The information relating to the XR500 R applies primarily to the US model, but can also be applied to any Canadian specification machines imported into the UK. In general, changes are confined to jet sizes and settings and these are covered in the specifications at the beginning of this Chapter.
17 The information covering the altitude and temperature compensation given for the XR250 R can also be applied to the XR500 R model. Note that the standard settings and jet sizes are as shown below and that the range of optional jets is smaller (see table). The method of calculating the compensation factor 'C' and its application is identical.

Standard settings:

Identification number	PD11B
Main jet (standard)	152
Jet needle setting	3rd groove
Pilot screw setting	$2\frac{1}{4}$ turns out

Optional main jets:

Size	Part Number
138	99101-357-138
140	99101-357-140
142	99101-357-142
145	99101-357-145

13 Air filter: modifications – all models

1 The XL250/500 R and R-C models are fitted with a revised filter element which is retained in the filter casing, behind the left-hand side panel, by a single central wingnut. This arrangement replaced the earlier type shown in Chapter 2 where the element was held in place by a plastic spring retainer.
2 In the case of the later XR250/500 R machines, the element is housed behind the right-hand side panel and retained by a short bracket. The bracket is slotted to allow the element to be pushed securely up against the casing, and the assembly is retained by a hose clip at the inlet side and by a single wingnut at the bracket.
3 With either type of element the cleaning method remains as described in Chapter 2, Section 10. When refitting the element, make sure that it seats fully and that no air can bypass it. This is particularly important when riding off-road; any dust entering the engine will cause rapid wear.

14 Fuel tank and tap: modifications – all models

1 The various models covered in this update Chapter have been the subject of detail cosmetic and styling changes over the years. Whilst this is normally confined to a change of paintwork or redesigned graphics, some modifications have a more tangible effect on the machine. In the case of the XL250/500 R and R-C machines, the tank is changed in shape, but the method of mounting, two rubber buffers at the front and a single rear retaining bolt, remain unaltered.
2 On the XR250/500 R models, however, the new tank shape also brought a change in the mounting method; the single rear bolt is replaced by two tank-mounted hooks onto which a rubber strap with a security cable is attached. The new mounting system allows the tank to be removed quickly, without using tools.
3 The XL models also featured a revised fuel tap. This is attached and operates in the same way as the previous type, but has a removable sediment bowl and filter on the underside of the tap. This feature allows any fine dirt or water droplets to be drained from the fuel system before they are drawn into the carburettor.

15 Ignition system: modifications – XL250/500 R and R-C models

1 The CDI ignition system used on previous XL models used a centrifugal automatic timing unit as a means of advancing the ignition as engine speed rises. This arrangement works well enough, and is retained on the XR models, but it does allow the possibility of wear in the mechanical timing unit, upsetting the accuracy of the system. The XL250/500 R and R-C models avoid this by dispensing with the mechanical advance arrangement.
2 The ignition pickup rotor assembly, unlike that of earlier models, in which the mechanical timing unit was integral, consists of the rotor only, and this is fixed in relation to the crankshaft. The ignition is advanced electronically by additional circuits in the CDI unit which sense engine speed by reading the waveform characteristics from the pickup, or pulser, coil. This information is used to control ignition advance up to a predetermined maximum advance point. The rest of the ignition system functions in the same way as earlier versions.

16 Testing the ignition source coil resistances: XL250/500 R and R-C models

1 Trace the alternator output leads back to the connector beneath the fuel tank. Separate the connector, and connect the positive (+) probe of a multimeter to the Black/red lead and the negative (–) probe to a sound earth (ground) point. Set the meter to the ohms scale and note the reading.

2 The specified resistance is 50 – 200 ohms. If a reading of zero ohms is shown it indicates a short circuit between the source coil and earth, whilst a reading of infinite resistance indicates a break in the windings. Either condition indicates the need for renewal of the source coil. Since this is an integral part of the alternator stator assembly, the entire unit will have to be renewed unless a local electrical specialist is able to undertake a repair. Failing this, try the local motorcycle breaker who may be able to supply a good used stator.

17 Testing the CDI unit resistances: XL250/500 R and R-C models

1 **Note:** Honda maintain that the correct resistance readings can only be obtained using the specified test equipment and ranges. These are as shown below:

Sanwa electric tester (part number 07308-0020000) x K ohms range
Kowa electric tester (TH-5H) x 1000 ohms range

If using meters of other makes or models, Honda advise that false readings may be obtained. If the correct equipment is not available, it may prove easier to take the CDI unit to a Honda dealer for testing, or to check by substituting a new unit.
2 Disconnect and remove the CDI unit from the frame, then check the resistances, using the connections shown in the accompanying table.

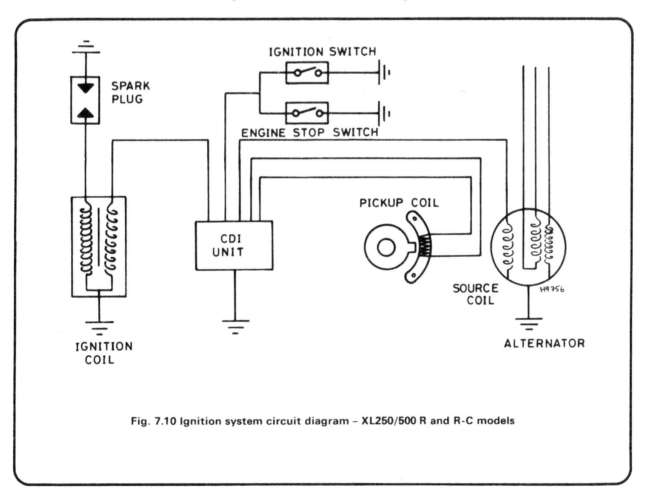

Fig. 7.10 Ignition system circuit diagram – XL250/500 R and R-C models

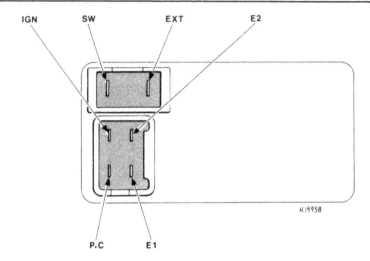

TEST RANGE: SANWA: x kΩ
KOWA: x 100Ω

+PROBE / -PROBE	SW	EXT	P●C	E1●E2	IGN
SW		∞	∞	∞	∞
EXT	0.1-20		*∞	*∞	∞
P●C	30-300	10-200		1-100	∞
E1●E2	1-50	0.1-20	1-100		∞
IGN	∞	∞	∞	∞	

Fig. 7.11 CDI unit test – UK XL250/500 R-C and US XL250/500 R models

Needle will swing and return to ∞

18 Testing the pickup coil resistance: XL250/500 R and R-C models

Trace the ignition pickup wiring back to the connector beneath the fuel tank and separate it. Measure the resistance between the Blue/Yellow lead and the Green/White lead. The correct resistance figure is 510 – 570 ohms. If the reading obtained is significantly different to this the pickup coil should be renewed.

19 Front fork modifications: general

XL250 S-B UK model

1   In the case of the UK XL250 S-B model, the fork design is changed to that shown for the XR models in Fig. 4.2. It follows that the fork specifications were altered accordingly, and these are shown at the start of this Chapter. When working on the forks refer to Chapter 4, Section 7, bearing in mind the new clearances and capacities.

XL250/500 R-C UK and XL250/500 R US models

2   The above-mentioned models employ another new fork assembly, again similar to the XR type. The twin spring arrangement used on the early models is replaced by single springs, with a plain washer and spacer fitted above. On all "R" models, the fork springs are supplemented by air valves, allowing pressurised air to be introduced at the top bolts. Damping oil quantity varies according to the model and production year, and these, together with the main fork specifications, will be found at the beginning of this Chapter. Dismantling and reassembly procedures are similar to those given in Chapter 4, but note the revised specifications. Details of the fork construction can be found in the accompanying line drawing.

XR250/500 R models

3   The XR250 R and XR500 R models feature revised forks with renewable bushes and air assistance. These are described in more detail in Section 20 of this Chapter and in the specifications at the beginning of the Chapter.

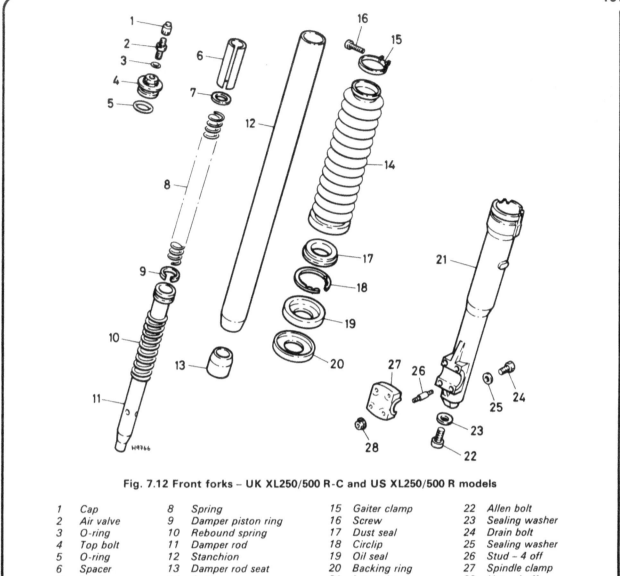

Fig. 7.12 Front forks – UK XL250/500 R-C and US XL250/500 R models

1	Cap	8	Spring	15	Gaiter clamp	22	Allen bolt
2	Air valve	9	Damper piston ring	16	Screw	23	Sealing washer
3	O-ring	10	Rebound spring	17	Dust seal	24	Drain bolt
4	Top bolt	11	Damper rod	18	Circlip	25	Sealing washer
5	O-ring	12	Stanchion	19	Oil seal	26	Stud – 4 off
6	Spacer	13	Damper rod seat	20	Backing ring	27	Spindle clamp
7	Washer	14	Gaiter	21	Lower leg	28	Nut – 4 off

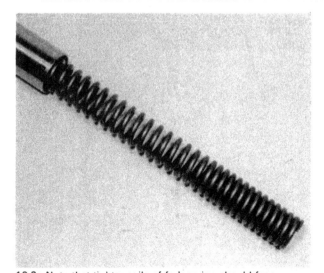

19.2a Note that tighter coils of fork spring should face uppermost

19.2b Do not omit to fit washer and spacer before fitting top bolt (XL250 and 500 R/R-C models)

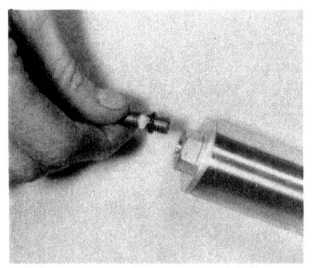

19.2c Schrader-type air valve screws into top bolt. Note O-ring seal

20 Front forks: removal and refitting – XR250 and 500 R models

1   The fork removal procedure is straightforward on the XR models and is much as described in Chapter 4. To summarise: remove the front wheel after disconnecting the speedometer and brake cables. Slacken the fork gaiter (boot) retaining clip immediately below the lower yoke. Slacken the upper and lower pinch bolts. The fork legs can now be pulled and twisted downwards until they come free of the yokes.

2   The fork oil should normally be changed whenever the fork legs are removed from the machine. Release the fork air pressure by depressing the air valve insert, then remove the top bolt from each leg. Invert the leg to drain it, pumping the fork to speed up the process. When adding oil to the fork legs, note the prescribed quantity and level figures given in the specifications. The fork oil level can be varied to give different suspension characteristics, and the exact level is largely a matter of personal preference.

3   Check the fork oil level by compressing the fork fully (spring removed). The oil level is the distance between the top edge of the stanchion and the surface of the oil with the fork leg held vertically. As a guide, the minimum level will give slightly stiffer

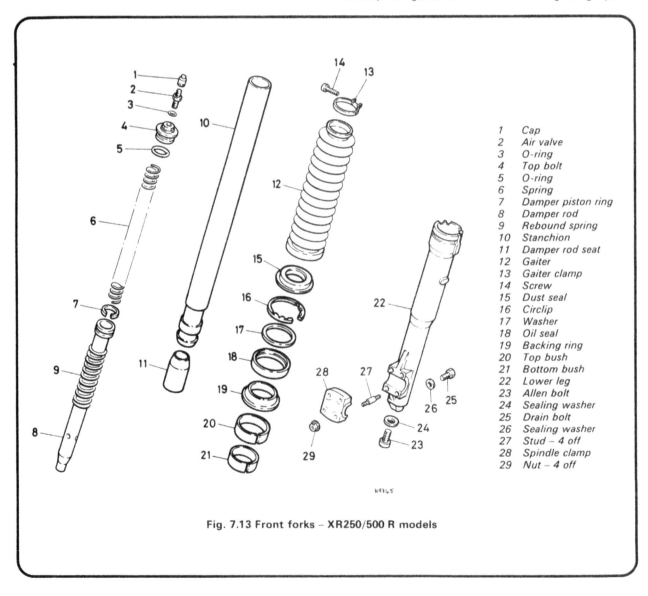

1	Cap
2	Air valve
3	O-ring
4	Top bolt
5	O-ring
6	Spring
7	Damper piston ring
8	Damper rod
9	Rebound spring
10	Stanchion
11	Damper rod seat
12	Gaiter
13	Gaiter clamp
14	Screw
15	Dust seal
16	Circlip
17	Washer
18	Oil seal
19	Backing ring
20	Top bush
21	Bottom bush
22	Lower leg
23	Allen bolt
24	Sealing washer
25	Drain bolt
26	Sealing washer
27	Stud – 4 off
28	Spindle clamp
29	Nut – 4 off

Fig. 7.13 Front forks – XR250/500 R models

springing due to the reduced air volume, whilst the maximum level will give the opposite effect. Check that the level is the same in each leg, and wipe the fork spring dry before refitting it.

4 When refitting the fork legs into the yokes, note the engraved line about 1 in down from the top of the stanchion. This should be aligned with the top surface of the upper yoke. Check that the fork top bolt and the pinch bolts are tightened to the specified torque figure.

5 Refer to Section 22 and set the fork air pressure to the recommended setting.

21 Front forks: dismantling and reassembly – XR250 and 500 R models

1 The procedure for dismantling the XR250/500 R type forks is broadly similar to that described in Chapter 4, Section 7, apart from the following differences. When separating the stanchion and lower leg, note that the forks are bushed. This means that the lower bush will not pass through the top bush, but this also provides a method for removing the latter from the lower leg. Make sure that the damper rod bolt is removed and that the circlip which retains the oil seal is released.

2 Pull the stanchion sharply outwards until the bushes contact, repeating this action until the top bush, backup ring and the oil seal are displaced. The stanchion assembly can then be withdrawn.

3 Before the forks are rebuilt, examine the sliding surfaces of both bushes. If either show a copper colour over more than $\frac{3}{4}$ of the width of the bush, renew them as a pair to maintain accurate fork action. The lower bush is split and can be worked past the shoulder on the stanchion. When fitting a new lower bush, avoid spreading it any more than is necessary to ease it into place. The stanchion assembly can now be lubricated and fitted into the lower leg and the damper rod bolt fitted.

4 To fit the top bush it is necessary to devise a method of driving it home squarely. Honda dealers can supply a tool for this purpose which consists of a tubular sliding weight. A similar home-made device is easily made up from scrap tubing with an internal diameter slightly greater than the stanchion. If making up a tool, remove all burrs carefully and wrap the stanchion with PVC tape to prevent scoring.

5 Lubricate the bush and slide it down over the stanchion, then fit the backup ring. Using the driver, tap the bush home until it seats fully and squarely. Check regularly that the bush is entering the bore square; if necessary remove it and start again. The driver tool can be used to press the lubricated oil seal into position. Do not omit to fit the circlip and the dust seal.

22 Front forks: setting the air pressure

1 The XL/XR250 and 500 R models (see Specifications) are fitted with air valves in the stanchion top caps. This is to permit the adjustment of the air pressure inside the fork leg to alter the fork's effective spring rate; the higher the air pressure, the stiffer will be the fork springing.

2 Two tools are essential for setting fork air pressure; a gauge capable of reading the low pressures involved, and a low-pressure pump. The gauge must be finely calibrated to ensure that both legs can be set to the same pressure, and must cause only a minimal drop in pressure whenever a reading is taken; as the total air volume is so small, an ordinary gauge, such as a tyre pressure gauge, will cause a large drop in pressure because of the amount of air required to operate it. Gauges for use on suspension components are now supplied by several companies and should be available through any good motorcycle dealer. The pump must be of the hand- or foot-operated type, a bicycle pump being ideal; aftermarket pumps are available for use on suspension systems and are very useful, but expensive. **Never** use a compressor-powered air line; it is all too easy to exceed

the maximum recommended pressure which may cause damage to the fork oil seals and may even result in personal injury. Add air very carefully, and in small amounts at a time.

3 The air pressue must be set when the forks are cold, and with the machine securely supported so that its front wheel is clear of the ground, thus ensuring that the air pressure is not artificially increased.

4 Set the pressure to the required amount within the specified range, then be careful to ensure that each leg is at **exactly** the same pressure. This is essential as any imbalance in pressures will impair fork performance and may render the machine unsafe to ride. Note that good quality aftermarket kits are now available to link separate air caps; the use of one of these, when correctly installed, will ensure that the pressures in the legs are equal at all times and will aid the task of setting the pressure in the future.

23 Steering head bearings: adjustment and overhaul – XL/XR250 and 500 R/R-C models

1 The later models feature taper roller steering head bearings in place of the cup-and-cone type used on the earlier machines. The new arrangement is considerably more robust than the earlier type and can be expected to last well, given regular adjustment and greasing.

2 Adjustment is carried out in the normal way, but note that the bearing adjuster nut must be tightened to a specified torque setting to apply the correct preload to the bearings. This is achieved using a special socket on the slotted adjuster nut, but most owners will be able to approximate the setting using a C-spanner. To set the correct torque figure, measure one foot along the handle of the C-spanner (extend it if required), from the centre of the C. Drill a hole at this point and use the hole to attach a simple spring balance. Tighten the nut by pulling on the spring balance until the prescribed pressure is obtained.

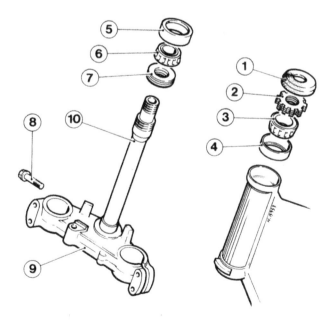

Fig. 7.14 Steering head – UK XL250/500 R-C and US XL/XR250 and 500 R models

1	Dust cover	6	Lower bearing race
2	Adjuster ring	7	Dust seal
3	Upper bearing race	8	Pinch bolt
4	Upper bearing shell	9	Lower yoke
5	Lower bearing shell	10	Steering stem

3   The steering head assembly can be dismantled for examination in the same way as described in Section 4 of Chapter 4. Under normal circumstances it will be necessary only to clean and regrease the bearings prior to reassembly, but if corrosion or wear necessitate the renewal of the bearings it should be noted that a press will be required to remove and fit the lower bearing. Owners having suitable facilities can carry out the work at home, otherwise have the job done by a Honda dealer.

24 Pro-link rear suspension: general

1   All of the R and R-C models covered in this update Chapter feature a rising-rate rear suspension system known as Honda Pro-Link. A conventional box-section swinging arm supported on needle roller bearings pivots from the frame in the usual manner, but is not connected directly to the single rear suspension unit. A linkage arrangement conveys movement from the swinging arm to the bottom of the suspension unit in such a way that the damping and springing effect is applied in a progressive manner in relation to swinging arm movement. Sintered metal bushes are used at the pivot points, these being sealed by renewable dust seals to prevent wear caused by the ingress of dirt.
2   If a small surface irregularity is encountered, the swinging arm and wheel are deflected by a small amount in response. During its initial travel, the swinging arm can move quite easily, with little compression of the suspension unit. As the amount of deflection at the wheel increases, so too does the rate at which the suspension unit is compressed. This has the effect of producing very soft and compliant initial movement, but this becomes progressively firmer as the amount of wheel deflection increases.
3   The result is that the wheel is able to track more accurately over various surfaces than would a conventional twin suspension unit arrangement with linear spring and damping rates. It is recommended that you ask someone to bounce up and down on the machine so that the action of the suspension linkage can be observed and understood more easily; it is far easier to grasp the rising rate effect by watching it in operation, than from a written description.
4   All models use a single central suspension unit comprising a damper unit containing oil and under nitrogen pressure, and a coil spring fitted around this. In the case of the XR models there is also an external reservoir connected to the damper unit, designed to prevent the damper unit from fading under competition conditions by allowing accumulated heat to be more readily dispersed. The XL models do not have the remote reservoir of the competition machines.

25 Rear suspension unit and linkage: removal and renovation – XL250/500 Pro-Link models

1   Support the machine securely so that the rear wheel is raised clear of the ground. This can be done by using a jack or stand beneath the engine, but make sure that the arrangement is stable to avoid any risk of the machine falling over. Remove both side panels, the seat and the rear mudguard to obtain access to the unit. Clean the swinging arm, linkage and suspension unit to remove any accumulated dirt before commencing dismantling work.
2   Working from below the swinging arm, remove the pivot bolt which secures the forked shock link to the frame. On the upper face of the swinging arm, loosen the shock arm pivot bolt and the suspension unit lower mounting bolt. Support the rear wheel with a jack or a wooden block, then remove both bolts to

free the lower end of the unit together with the linkage. Slacken the suspension unit upper mounting bolt, then supporting the unit with one hand, remove the bolt to free it from the frame. The suspension unit and the linkage can now be manoeuvred clear of the frame and placed to one side to await examination.
3   It should be noted at this stage that it is not possible to dismantle the damper unit, although the spring can be removed. To release the spring, clamp the unit upside down in a vice using a rag or soft jaws to prevent damage. Using C-spanners, slacken the locknut and adjuster nut to release spring tension. Run the slotted nuts off the damper unit then lift the spring away. If required, hold the lower mounting in the vice and slacken the locknut. The lower mounting can now be unscrewed and the dust seal, spring guide, spring seat and spring seat stopper removed.
4   Clean the damper body, noting any signs of oil leakage. If the damper is faulty it must be renewed; it is not possible to dismantle or repair it. If the unit is leaking it is likely that it will have lost gas pressure, and this can be checked by noting the pressure required to compress the damper unit (less spring). Place one end of the unit on some bathroom scales and note the reading shown when the damper is depressed. If less than the specified minimum figure, the unit is leaking and should be renewed. Renewal will also be required if the damper rod is scored, corroded or bent. Before reassembling the unit, check that the spring free length is within limits. If less than the minimum figure, fit a new spring.

Fig. 7.15 Method of testing suspension unit damper – Pro-link models

5   Check the various linkage pivots for signs of wear or damage, and renew the bushes as required. The bushes can be removed and fitted using a vice and sockets as a simple press as shown in the accompanying photographs. If wear is discovered it is probably a result of inadequate lubrication or failed seals. It is a good idea to renew the seals as a matter of course to ensure that dirt cannot enter the pivots in the future. When assembling the pivots, lubricate them with molybdenum

disulphide paste. Honda recommend Molykote G-n Paste, Locol Paste or a suitable equivalent. If in doubt, consult a Honda dealer who should be able to recommend an alternative product. Reassemble the linkage components, tightening the pivot bolts to the specified torque settings.

6   Before reassembly of the rear suspension unit commences, check that all parts are clean. Fit the dust seal, spring guide, spring seat and spring seat stopper. Fit the lower mounting, using Loctite or similar on the threads and ensuring that the pin on the spring seat stopper aligns with the notch in the lower mounting. Tighten the locknut to 6.0 – 7.5 kgf m (45 – 54 lbf ft). Fit the spring and adjuster nut and set the spring free length to 241 mm (9.49 in) in the case of the 500 model and 246.5 mm (9.7 in) in the case of the 250 to give the correct preload. Secure the locknut to hold this setting.

7   Lubricate the suspension unit pivots with molybdenum disulphide paste (see above) and then install the unit by reversing the dismantling sequence. Tighten all pivots to the prescribed torque figure, and check that the suspension operates normally before refitting the mudguard, seat and side panels.

25.2a Loosen shock arm pivot bolt and suspension unit lower mounting bolt

25.2b Suspension unit upper mounting bolt can now be removed to free unit and linkage

25.5a Using a vice and sockets to press out linkage bushes

25.5b Clean shock link assembly ...

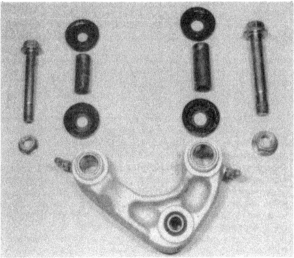

25.5c ... and shock arm assembly – examine for wear or damage

25.5d Do not forget the suspension unit bushes

26 Rear suspension unit and linkage: removal and renovation – XR250/500 Pro-Link models

1   Support the machine securely so that the rear wheel is raised clear of the ground. This can be done by using a jack or stand beneath the engine, but make sure that the arrangement is stable to avoid any risk of the machine falling over. Remove both side panels, the seat, the exhaust system and the air filter casing, plugging the carburettor inlet with rag to exclude dirt. Clean the swinging arm, linkage and suspension unit to remove any accumulated dirt before commencing dismantling work.

2   Release the clamp which holds the suspension unit reservoir in position on the frame. Remove the suspension unit upper mounting bolt, and slacken, but do not remove, the suspension unit lower mounting bolt. Moving to the underside of the swinging arm, remove the bolt securing the shock link to the frame. **Do not** remove the shock link to shock arm bolt. Tip the suspension unit back until it is vertical, then raise the rear wheel so that the suspension unit lower mounting bolt is clear of the swinging arm. Support the wheel in this position by placing a block beneath it. The lower mounting bolt can now be removed and the unit and reservoir lifted away, leaving the linkage in place.

3   It should be noted at this stage that it is not possible to dismantle the damper unit, though the spring can be removed and the oil can be drained and changed. To release the spring, clamp the unit upside down in a vice using a rag or soft jaws to prevent damage. Using C-spanners, slacken the locknut and adjuster nut to release spring tension.

4   In the case of the XR models the reservoir hose prevents the spring being removed from the top so provision is made to allow its removal over the lower mounting. Press the spring seat in against the spring and disengage the C-shaped retainer. The spring seat and the spring can now be lifted away. Although it is theoretically possible to drain and change the damping oil it should be noted that the unit must be depressurised before the reservoir is removed and that during reassembly it is necessary to add nitrogen to a pressure of 20 kg/cm² (285 psi). It is unlikely that the average owner will have access to the necessary equipment, and it is strongly recommended that such work is entrusted to a Honda dealer.

5   Clean the damper body, noting any signs of oil leakage. If the damper is faulty, it must be renewed; it is not possible to dismantle or repair it. If the unit is leaking it is likely that it will have lost gas pressure, and this can be checked by noting the

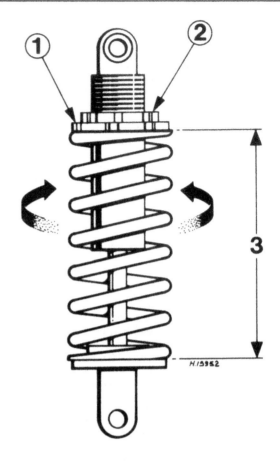

Fig. 7.16 Spring free length measurement – Pro-link models

| 1 | Adjuster nut | 3 | Free length |
| 2 | Locknut | | |

pressure required to compress the damper unit (less spring). Place one end of the unit on some bathroom scales and note the reading shown when the damper is depressed. If less than the specified minimum figure, the unit is leaking and should be renewed. Renewal will also be required if the damper rod is scored, corroded or bent. Before reassembling the unit, check that the spring free length is within limits. If less than the minimum figure, fit a new spring.

6   Check that all parts are clean before reassembly commences. Fit the spring, spring seat and the C-shaped retainer. Fit the spring and adjuster nut and set the spring free length to 212.6 mm (8.4 in) to give the correct preload. The spring preload can be varied to suit personal preference by altering the effective length of the spring. Note that on no account must it vary from the standard setting by more than 5 mm (¼ in) or damage may result. Each turn of the adjuster nut will alter the effective spring length by 1.5 mm (0.06 in). When adjustment has been completed, fit and secure the locknut.

7   It is a good idea to check the linkage pivots while the suspension unit is removed. Look for signs of wear or damage, and renew the bushes as required. The bushes can be removed and fitted using a vice and sockets as a simple press as shown in the photographs which accompany the previous section. If wear is discovered it is probably a result of inadequate lubrication or failed seals. It is good practice to renew the seals as a matter of course to ensure that dirt cannot enter the pivots

in the future. When assembling the pivots, lubricate them with molybdenum disulphide paste. Honda recommend Molykote G-n Paste, Locol Paste or a suitable equivalent. If in doubt, consult a Honda dealer who should be able to recommend an alternative product. Reassemble the linkage components, tightening the pivot bolts to the specified torque settings.

8   Lubricate the suspension unit pivots with molybdenum disulphide paste (see above) and then install the unit by reversing the dismantling sequence. Tighten all pivots to the prescribed torque figure, and check that the suspension operates normally before refitting the exhaust, air filter, mudguard, seat and side panels.

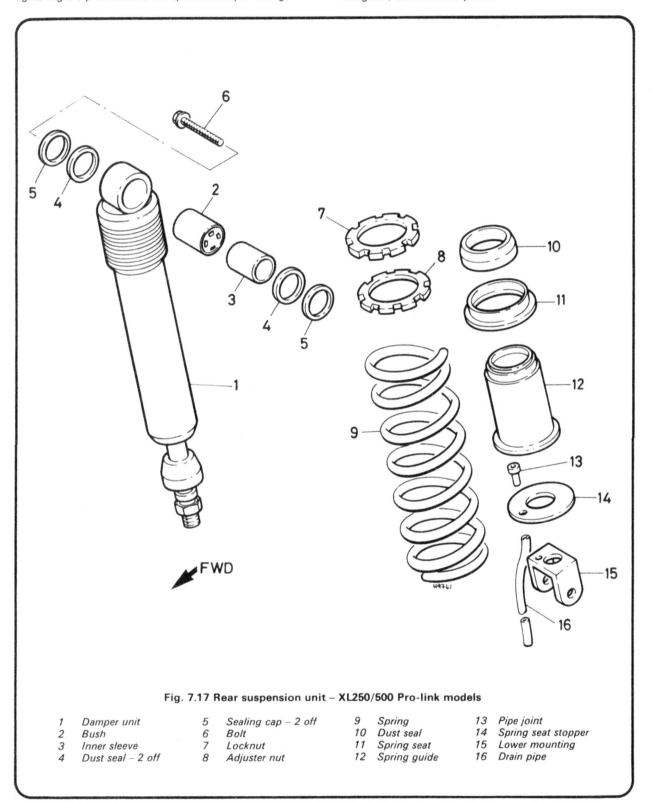

Fig. 7.17 Rear suspension unit – XL250/500 Pro-link models

1	Damper unit	5	Sealing cap – 2 off	9	Spring	13	Pipe joint
2	Bush	6	Bolt	10	Dust seal	14	Spring seat stopper
3	Inner sleeve	7	Locknut	11	Spring seat	15	Lower mounting
4	Dust seal – 2 off	8	Adjuster nut	12	Spring guide	16	Drain pipe

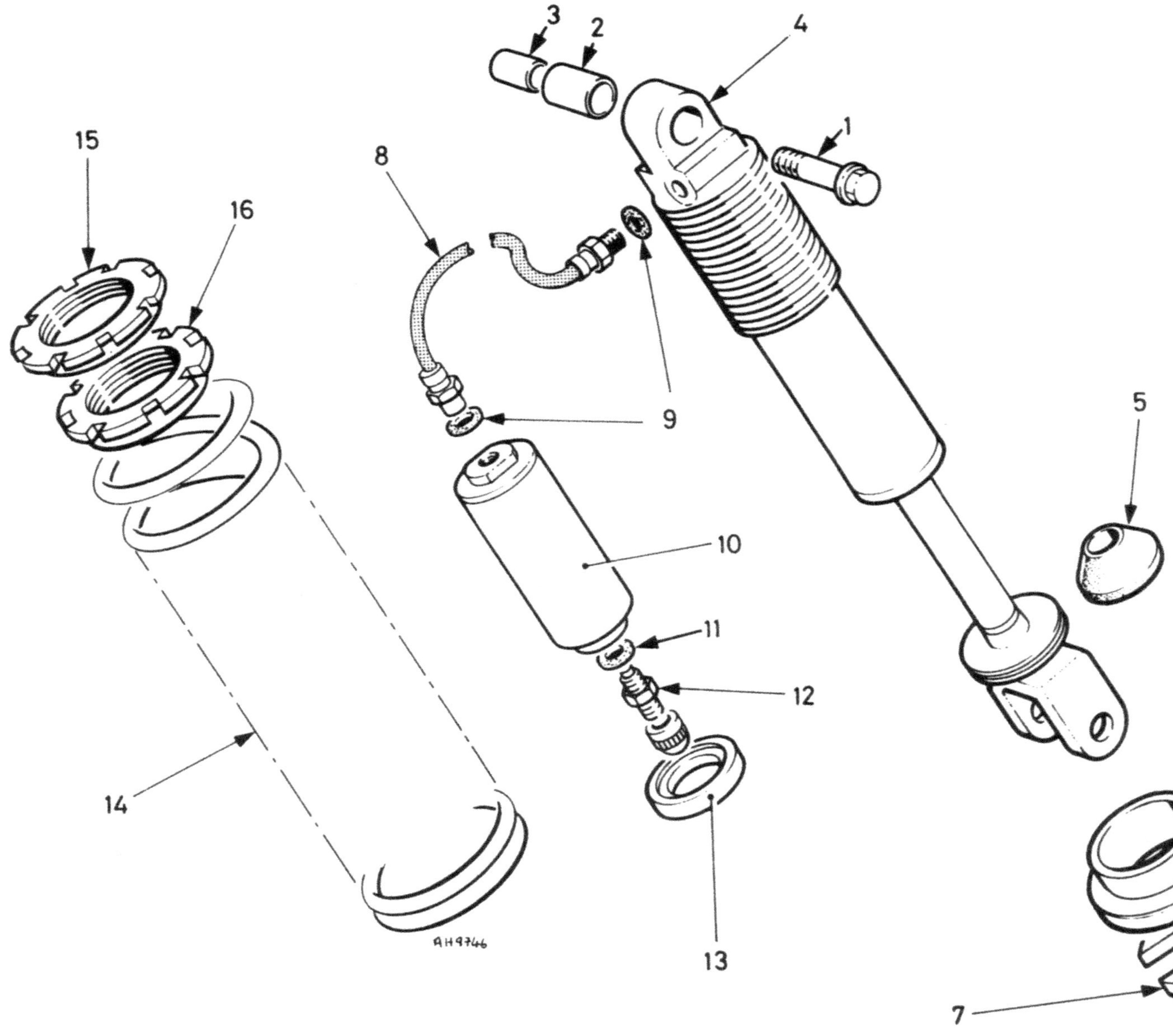

Fig. 7.18 Rear suspension unit –
XR250/500 Pro-link models

1   Bolt
2   Bush
3   Inner sleeve
4   Damper unit
5   Bump stop
6   Spring seat
7   C-shaped retainer
8   Reservoir hose
9   O-ring – 2 off
10 Reservoir
11 O-ring
12 Air valve
13 Cap
14 Spring
15 Locknut
16 Adjuster nut

AH9766

Fig. 7.19 Suspension linkage – Pro-link models

1 Forked shock link
2 Pivot bolt
3 End cap – 2 off
4 Bush
5 Inner sleeve
6 Grease nipple
7 Shock arm
8 Pivot bolt – 2 off
9 End cap – 4 off
10 Bush – 2 off
11 Inner sleeve – 2 off
12 Washer – 2 off
13 Nut – 2 off
14 Grease nipple – 2 off
15 Pivot bolt
16 Bush

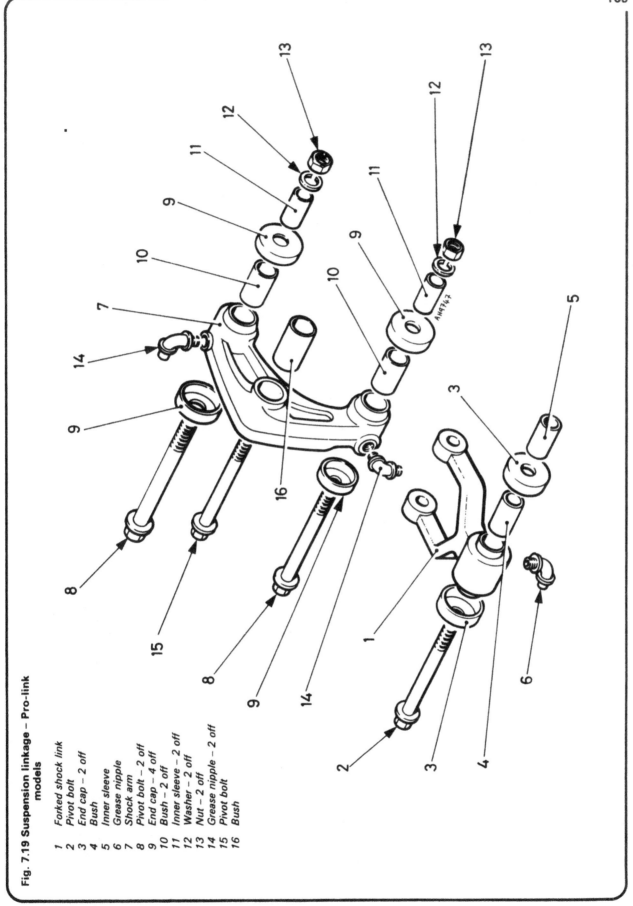

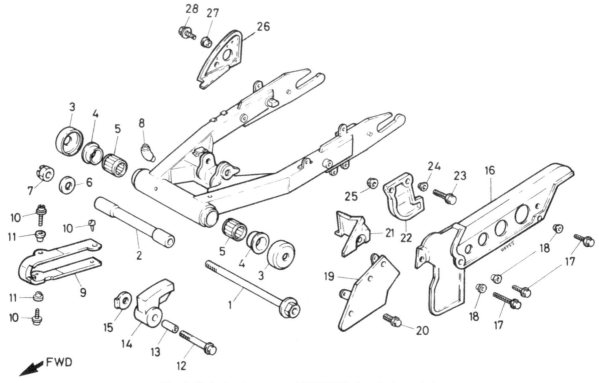

Fig. 7.20 Swinging arm – XL250/500 Pro-link models

1	Pivot bolt	8	Grease nipple
2	Centre sleeve	9	Chain guide
3	Dust cap – 2 off	10	Screw – 3 off
4	Headed thrust bush	11	Spacer – 2 off
5	Bearing – 2 off	12	Bolt
6	Washer	13	Spacer
7	Nut	14	Chain guide

15	Nut	22	Foot guard
16	Chainguard	23	Bolt
17	Bolt – 3 off	24	Spacer
18	Spacer – 3 off	25	Nut
19	Mudguard	26	Foot guard
20	Bolt	27	Spacer
21	Mudguard	28	Bolt

27 Swinging arm: removal and renovation – XL250/500 Pro-Link models

1　Support the machine securely so that the rear wheel is raised clear of the ground. This can be done by using a jack or stand beneath the engine, but make sure that the arrangement is stable to avoid any risk of the machine falling over. Remove both side panels, the seat, and the rear mudguard. Clean the swinging arm, linkage and suspension unit to remove any accumulated dirt before commencing dismantling work.

2　Remove the rear wheel, the chainguard and the rear brake return spring. Free the lower end of the rear suspension unit by removing the pivot bolt which retains it. Slacken the 12 mm engine mounting bolt located just above the swinging arm pivot, then remove the pivot nut. Support the swinging arm and displace and remove the pivot bolt to allow it to be lifted clear of the frame. Remove as necessary the suspension linkage, the rubber chain guide, the passenger footrests and the foot guard.

3　Clean and dismantle the suspension linkage, checking the pivot bushes as described in section 25. The swinging arm pivot bearings must not be removed other than to renew them; they will be damaged during removal. To extract the bearings a proprietary bearing removal tool will be necessary (see photograph). Few owners will own such a tool, but it may be possible to hire this equipment locally. Failing this, it is probably best to entrust the job to a Honda dealer.

4　The new bearings can be fitted using a double-diameter drift to prevent distortion. Grease the new bearing thoroughly and tap it into place together with its headed thrust bush. It

should be noted that the marked face of the bearing must face outwards. Refit the swinging arm by reversing the removal sequence, tightening all fasteners to the prescribed torque setting.

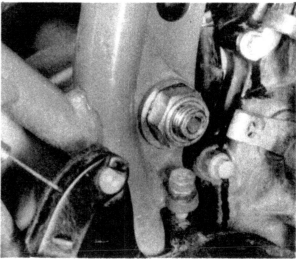

27.2a Slacken and remove the swinging arm pivot nut ...

27.2b ... and withdraw the spindle to free the swinging arm assembly

27.2c Chain roller assembly should be cleaned and its pivots lubricated

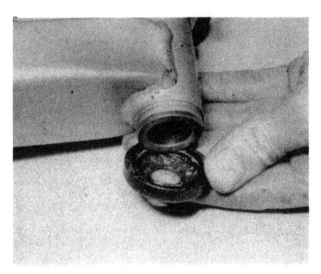

27.3a Remove seals from ends of the swinging arm ...

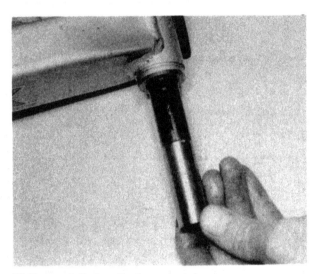

27.3b ... and displace the inner pivot sleeve ...

27.3c ... to gain access to needle roller pivot bearings

27.3d Bearings should be removed using a proprietary extractor

27.4 Remember to grease the linkage as shown after reassembly

28 Swinging arm: removal and renovation – XR250/500 Pro-Link models

1   Support the machine securely so that the rear wheel is raised clear of the ground. This can be done by using a jack or stand beneath the engine, but make sure that the arrangement is stable to avoid any risk of the machine falling over. Remove the seat, the air filter casing and the exhaust system. Clean the swinging arm, linkage and suspension unit to remove any accumulated dirt before commencing dismantling work.

2   Remove the rear wheel, and the rear brake cable. Free the lower end of the rear suspension unit by removing the pivot bolt which retains it. Remove the bolt which retains the shock link to the frame. Slacken and remove the swinging arm pivot nut. Support the swinging arm and displace and remove the pivot bolt to allow it to be lifted clear of the frame.

3   It is also possible to remove the swinging arm together with the rear suspension unit, in which case the air filter casing and exhaust system need not be disturbed. If this approach is chosen, follow the sequence shown below.

a)   Remove the seat
b)   Remove the rear wheel
c)   Remove the suspension unit upper mounting bolt and reservoir
d)   Remove the shock link to frame bolt
e)   Remove the swinging arm pivot bolt
f)   Lower the swinging arm, linkage and suspension unit clear of the frame.

4   Clean and dismantle the suspension linkage, checking the pivot bushes as described in Section 26 above. The swinging arm pivot bearings must not be removed other than to renew them; they will be damaged during removal. To extract the bearings a proprietary bearing removal tool will be necessary (see photograph). Few owners will own such a tool, but it may be possible to hire this equipment locally. Failing this, it is probably best to entrust the job to a Honda dealer.

5   The new bearings can be fitted using a double-diameter drift to prevent distortion. Grease the new bearing thoroughly

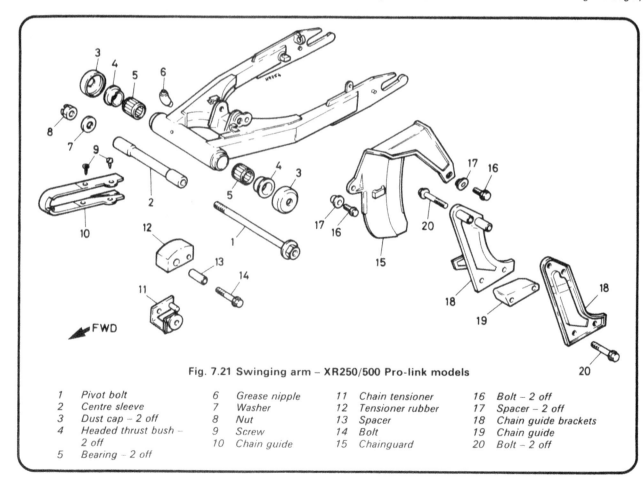

Fig. 7.21 Swinging arm – XR250/500 Pro-link models

1	Pivot bolt	6	Grease nipple	11	Chain tensioner	16	Bolt – 2 off
2	Centre sleeve	7	Washer	12	Tensioner rubber	17	Spacer – 2 off
3	Dust cap – 2 off	8	Nut	13	Spacer	18	Chain guide brackets
4	Headed thrust bush – 2 off	9	Screw	14	Bolt	19	Chain guide
5	Bearing – 2 off	10	Chain guide	15	Chainguard	20	Bolt – 2 off

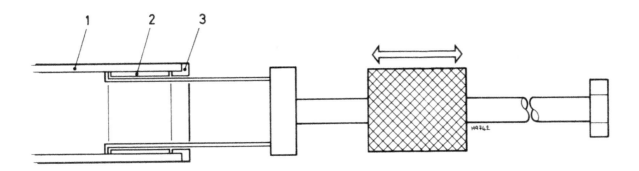

Fig. 7.22 Recommended method of swinging arm bearing removal

1   Swinging arm   2   Bearing   3   Headed thrust bush

and tap it into place together with its headed thrust bush. It should be noted that the marked face of the bearing must face outwards. Check the condition of the rubber chain guide and the nylon guide roller. Refit the swinging arm by reversing the removal sequence, tightening all fasteners to the prescribed torque setting.

29 Front brake: general – XL500 R, R-C and XR250/500 R models

1   With the exception of the XL250 R and R-C models, all 'R' versions covered by this update Chapter were equipped with a twin leading shoe (tls) front brake. In this type of brake there are two operating arms and brake cams, linked by a short adjustable rod. This allows both shoes to "lead"; that is, the moving or cam end of each shoe is brought into contact with the oncoming drum surface. This creates a self-servo effect which increases braking effort; the rotating drum surface tends to drag the brake shoe into firmer contact with the drum than lever pressure alone would suggest. In a single leading shoe (sls) drum brake, one shoe is always trailing and is thus only fully effective if the machine is moving backwards.

2   Whilst the tls brake allows more braking effort from a given size of brake drum it is not without its own disadvantages. Firstly, it is more complex and requires careful setting up to ensure that it is operating to full efficiency, particularly when new brake shoes have been fitted. Secondly, there is the aforementioned lack of braking power when the wheel is turning backwards. Whilst this is normally of little significance on the road, it should be remembered that extra lever pressure may be needed when checking the machine from rolling back down a slope when riding off-road.

30 Twin leading shoe front brake: examination and renovation – XL500 R, R-C and XR250/500 R models

1   With the front wheel removed from the machine the brake backplate assembly can be lifted away for examination. Before dismantling commences, mark the brake shoes with a spirit-based felt marker so that they can be refitted in their original positions. This is important if the old shoes are to be reused, but

if they are to be renewed this precaution can be ignored. Remove the shoes by pulling them away from their pivots and cams and folding them inwards until return spring pressure is released and they can be lifted away. Note that in the case of the XR models the shoe ends are located by plates secured by split pins; these must be removed before the shoes can be released.

2   Remove the pinch bolt which secures each brake arm to its spline. **Do not** disturb the connecting rod or synchronisation will be lost. Lift away the brake arm and connecting rod as an assembly. Displace the brake cams from the brake backplate. Clean and degrease the brake components (including the drum) using a solvent. Never use compressed air and take great care to avoid breathing brake dust; it contains asbestos and is toxic. Use a non-greasy solvent such as methylated spirit or petrol, but take adequate fire precautions and work outside, or in a well ventilated area, away from any fire risk.

3   The nominal thickness of the brake linings is 4.0 mm (0.16 in) and they should be renewed as a pair if worn to 2.0 mm (0.08 in) at any point. Never renew the shoes single. Each shoe will show most wear at the cam end, but wear should be similar on each; if there is a significant difference it indicates poor synchronisation between the brake arms and this should be checked during reassembly. Check the brake drum for scoring or damage. Light scoring can be removed using abrasive paper. Deeper scores may be corrected by skimming in a lathe, unless this means that the maximum diameter of the drum is exceeded. The drum maximum diameter on the 500 models is 131.0 mm (5.16 in) whilst in the case of the XR250 R it is 141.0 mm (5.55 in). If badly worn or distorted, have the wheel rim built onto a new hub and drum unit.

4   During assembly take care to keep grease away from the lining and drum surfaces. To this end, cover the lining surfaces with masking tape until just before the brake backplate is fitted into the drum, but do remember to remove it! Fit the seat washers and inner seals (XL only) over the brake cams, grease the cam spindles and insert them in the brake backplate. Refit the outer seals. Note that the lower cam carries the longer brake arm and should be fitted with a dust seal, plain washer and return spring (XR only), whilst the remaining cam should be fitted with a dust seal, return spring (XL only) and wear indicator. Fit the brake shoes by reversing the removal procedure, having first greased the shoe ends sparingly. On XR models, refit the retainer plates and secure using new split pins.

5   Offer up the brake arms and connecting rod as an assembly, again taking care not to disturb the rod position. Check that the arms are fitted so that the alignment marks coincide with those on the ends of the cam spindles. Tighten the upper (short) arm pinch bolt to 0.8 – 1.2 kgf m (6 – 9 lbf ft) and the lower arm

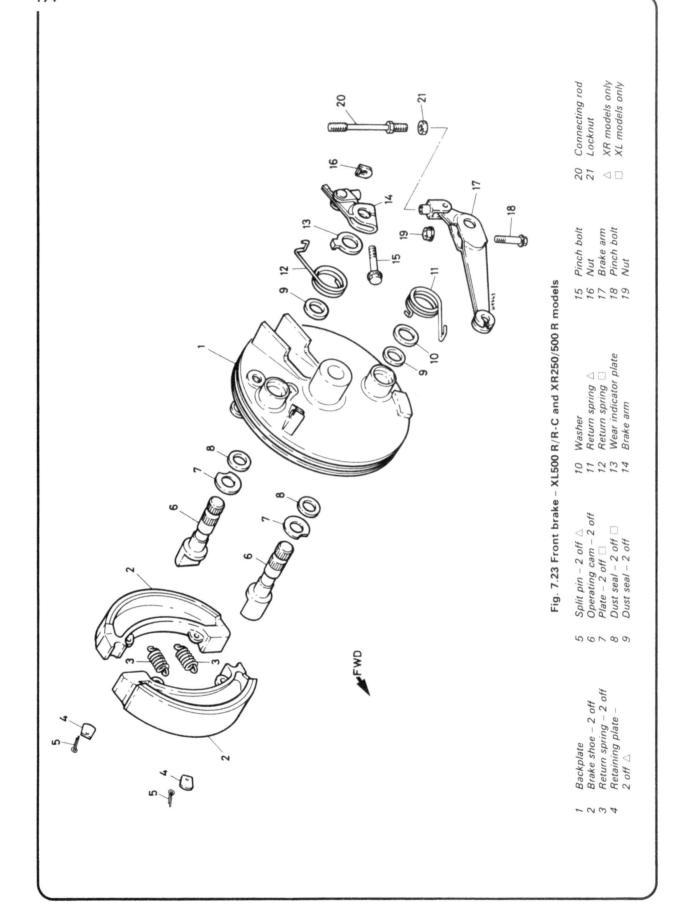

Fig. 7.23 Front brake – XL500 R/R-C and XR250/500 R models

1 Backplate
2 Brake shoe – 2 off
3 Return spring – 2 off
4 Retaining plate –
   2 off △
5 Split pin – 2 off △
6 Operating cam – 2 off
7 Plate – 2 off □
8 Dust seal – 2 off □
9 Dust seal – 2 off
10 Washer
11 Return spring △
12 Return spring □
13 Wear indicator plate
14 Brake arm
15 Pinch bolt
16 Nut
17 Brake arm
18 Pinch bolt
19 Nut
20 Connecting rod
21 Locknut
△ XR models only
□ XL models only

FWD

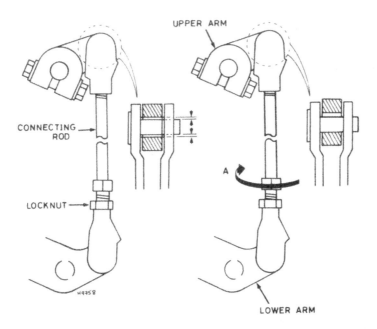

Fig. 7.24 Front brake cam synchronisation – XL500 R/R-C and XR250/500 R models

pinch bolt to 1.0 – 1.4 kgf m (7 – 10 lbf ft). If the arm synchronisation was not disturbed and is known to be accurate, the brake can now be refitted, otherwise proceed as described below.

6   The synchronisation between the brake arms and cams should be checked if the connecting rod or a cam have been renewed or where there is any doubt as to its accuracy. Poor synchronisation can have a profound effect on braking performance and should not be ignored. Start by slackening the connecting rod locknut. Grasp the shoes in one hand and squeeze them together to remove any play between the shoe ends and the cams. Turn the connecting rod until free play can be felt in the linkage, then gradually turn the rod in the direction indicated by the arrow (A) in the accompanying line drawing until the free play is just removed Do not move the rod any further than this. The purpose of the adjustment is to remove all free play, but without forcing the cams out of parallel. Secure the locknut and check that the cams lie parallel and that both start to move as soon as the main brake arm is operated. Once adjustment is checked, the brake assembly and wheel can be refitted.

31 Rear wheel modifications: general – XL250/500 R, R-C and XR250/500 R models

There are a number of detail changes to the rear wheel and brake of the Pro-Link models. These mostly relate to the change in the suspension system and will be self explanatory during examination. The most significant alteration is to the chain tension adjuster which is of the snail cam type. To adjust chain tension, slacken the rear wheel spindle and turn both cams by an equal amount to alter the tension. Check that the alignment marks on each side are the same to ensure wheel alignment. The drive chain free play figures are as follows:

XL models: 30 – 40 mm ($1\frac{1}{4}$ – $1\frac{5}{8}$ in) measured at the centre of the lower run with the machine on its side stand
XR models: 35 – 45 mm ($1\frac{3}{8}$ – $1\frac{3}{4}$ in) measured at the centre of the upper run with the rear wheel supported clear of the ground

31.1 Numbers on snail cam adjusters allow wheel alignment to be maintained

32 Electrical system modifications: general – XL250/500 R, R-C and XR250/500 R models

The electrical systems of the later models received varying degrees of attention and modification. In the case of the XR machines, with only a rudimentary system for lighting, this is confined to a change in the alternator wattage from 47W to 50W. Where the XL models are concerned things are rather different, with a complete new 12 volt electrical system in place of the previous 6 volt arrangement. Other changes were the result of cosmetic alterations to the machines and are mostly self explanatory. The sections which follow relate to the more significant alterations to the XL250/500 R and R-C machines.

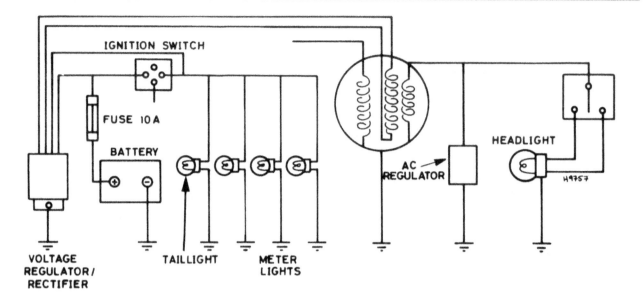

Fig. 7.25 Charging system circuit diagram – XL500 R and R-C models

33 Checking the charging output: XL250/500 R and R-C models

1 This test can be made using a voltmeter (or multimeter set on the appropriate voltage range) and an ammeter. Before commencing the test make sure that the engine is at its normal operating temperature and that the battery is fully charged. Trace the wiring from the regulator/rectifier unit and disconnect the black lead only. Connect the ammeter and voltmeter as shown in the accompanying illustration.

2 Start the engine and gradually increase the engine speed, noting the readings shown on the two meters. Compare these with those shown for the model being tested in the specifications at the beginning of this Chapter.

3 If no charging output is shown, check that all connections and wiring are in good condition. If this fails to resolve the problem, check the alternator windings as described below.

34 Alternator winding resistance checks: XL250/500 R and R-C models

1 The condition of the alternator stator windings can be checked with the alternator in position using a multimeter set on the ohms x 1 scale. Trace the alternator output wiring back to the block connector and separate it. To check the charging coil, test the resistance between the Pink and Yellow leads. To check the lighting coil, measure the resistance between the White/Yellow lead and earth (ground).

2 In each case a reading of 0.2 – 1.0 ohms should be indicated. If infinite resistance is shown an internal break in the stator windings is indicated, necessitating the renewal of the stator.

35 Voltage regulator/rectifier unit resistance test: XL250/500 R and R-C models

1 To test the regulator/rectifier unit it is necessary to make resistance tests using one of two test meters specified by Honda. In the absence of the correct meter, accurate readings

cannot be guaranteed, and the testing is best entrusted to a Honda dealer. The specified meters are the Sanwa electrical tester, part number 07308-0020000 or the Kowa electrical tester (TH-5H). In the case of the Sanwa tester, set the meter to the x K ohms scale, and on the Kowa tester select the ohms x 100 scale. In either case, ensure that the meter has good batteries and is correctly zeroed for the range selected.

2 Trace and disconnect the regulator/rectifier leads at the block connector. With the meter set on the appropriate range, check the resistances according to the accompanying table. If one or more resistance readings is other than specified, the unit must be renewed; it is not possible to repair it.

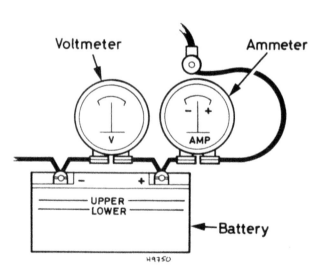

Fig. 7.26 Charging output test – XL250/500 R and R-C models

34.2a Alternator stator assembly is secured by three bolts ...

34.2b ... to support bracket inside the outer cover

G	P	
	Y	
Bl	R	

H9769

RANGE : SANWA : kΩ
KOWA : x100Ω

+PROBE / −PROBE	YELLOW	PINK	GREEN	RED	BLACK
YELLOW		∞	∞	1−20	∞
PINK	∞		∞	1−20	∞
GREEN	1−20	1−20		3−100	0.2−20
RED	∞	∞	∞		∞
BLACK	1−50	1−50	0.2−10	3−100	

H.15969

**Fig. 7.27 Regulator/rectifier unit test table –
XL250/500 R and R-C models**

36 AC regulator test: XL250/500 R and R-C models

1   Remove the headlamp unit and connect a voltmeter (or multimeter set to the appropriate range) as shown in the accompanying illustration. Switch on the lights and select main (Hi) beam on the dipswitch. With the engine running, note the reading at 5000 rpm. If this is outside the range 13.5 − 14.5 volts, check the AC regulator resistances as described below.

2   Note that the Honda specified meters and settings mentioned in Section 35 must be used for this test. Trace the AC regulator leads back to their bullet connectors and separate them. Connect the multimeter probes to the White and the Green leads and note the reading. Now reverse the probes and again note the reading. In each case a reading of 10 − 900 ohms should be shown. If outside this range the unit is defective and must be renewed. In the absence of the prescribed test meters, check by substitution, or have the test carried out by a Honda dealer.

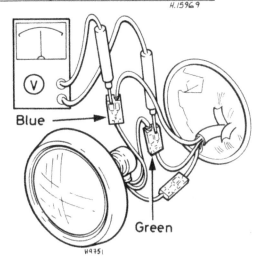

Blue
Green

H9751

**Fig. 7.28 AC regulator test – XL250/500 R and R-C
models**

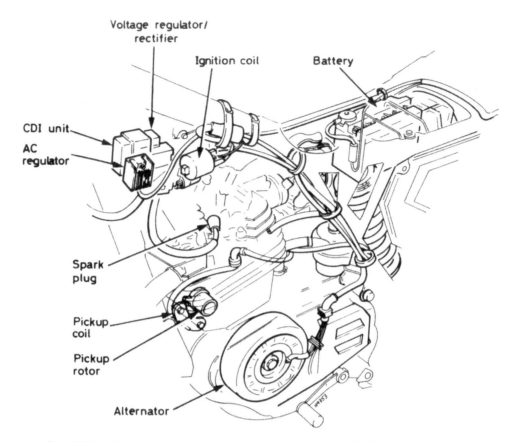

Fig. 7.29 Ignition and electrical component locations – XL250 R and R-C models

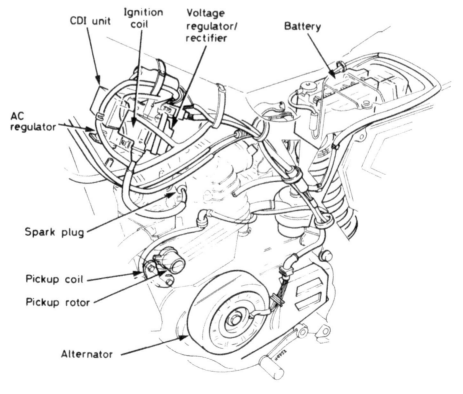

Fig. 7.30 Ignition and electrical component locations – XL500 R and R-C models

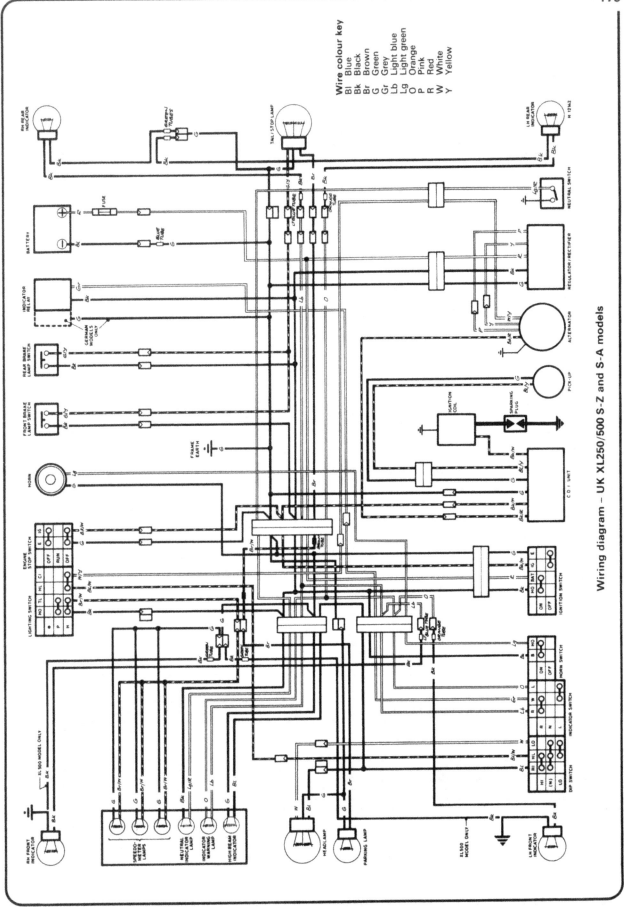

Wiring diagram – UK XL250/500 S-Z and S-A models

Wire colour key

Bl Blue
Bk Black
Br Brown
G Green
Gr Grey
Lb Light blue
Lg Light green
O Orange
P Pink
R Red
W White
Y Yellow

Wiring diagram – UK XL250 S - B model

Wire colour key

B	Blue
Bk	Black
Br	Brown
G	Green
Gr	Grey
Lb	Light blue
Lg	Light green
O	Orange
P	Pink
R	Red
W	White
Y	Yellow

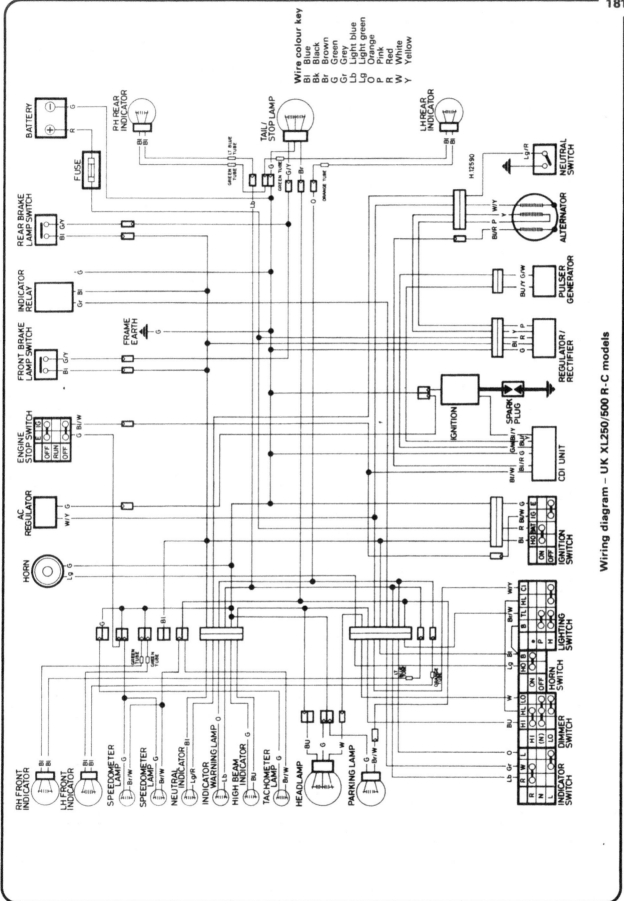

Wiring diagram – UK XL250/500 R-C models

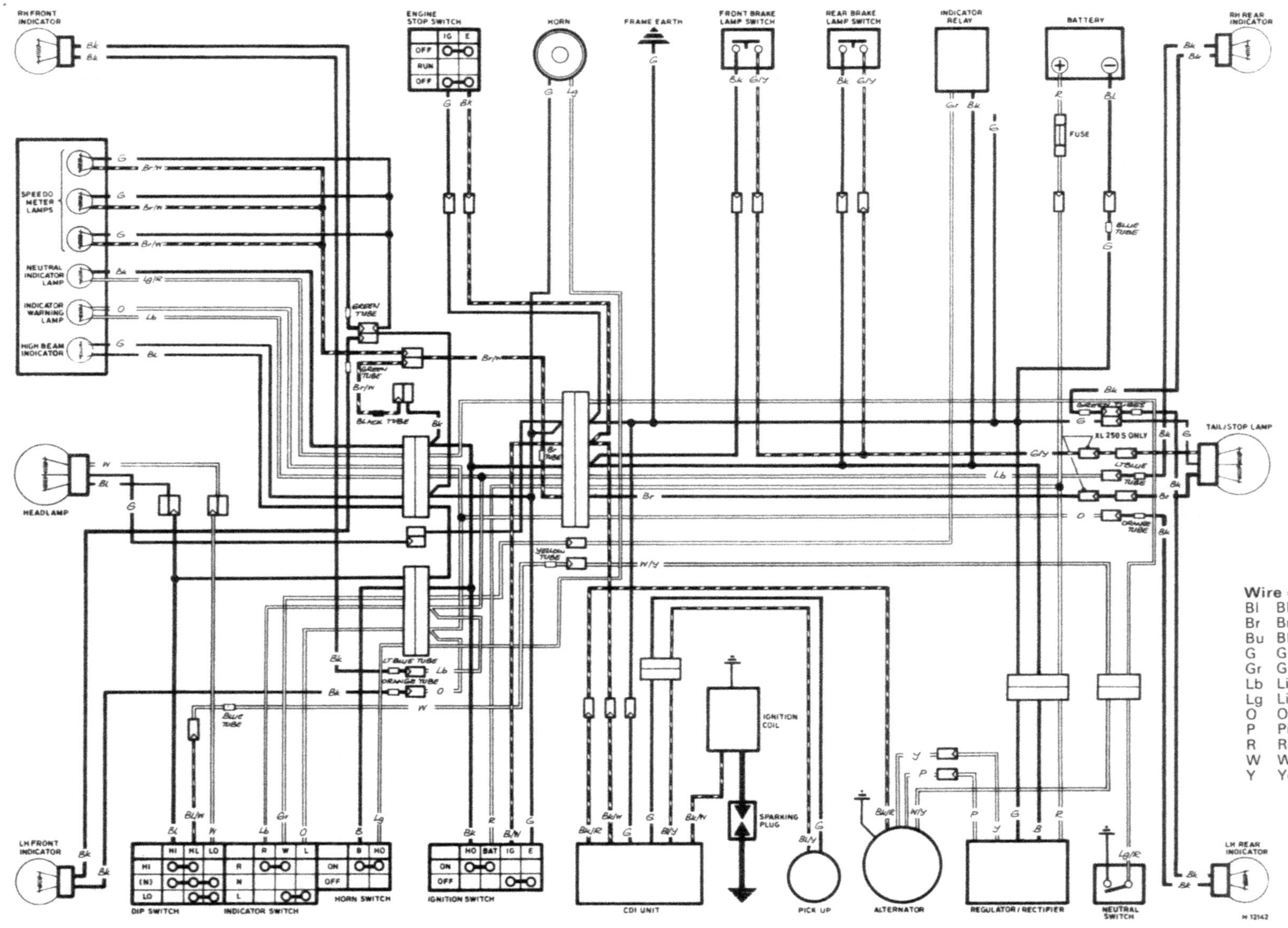

Wire colour key
Bl Black
Br Brown
Bu Blue
G   Green
Gr Grey
Lb Light blue
Lg Light green
O   Orange
P   Pink
R   Red
W   White
Y   Yellow

RH REAR INDICATOR

TAIL/STOP LAMP

LH REAR INDICATOR

BATTERY

FUSE

INDICATOR RELAY

REAR BRAKE LAMP SWITCH

FRONT BRAKE LAMP SWITCH

FRAME EARTH

HORN

ENGINE STOP SWITCH

NEUTRAL SWITCH

REGULATOR/RECTIFIER

ALTERNATOR

PICK UP

IGNITION COIL

SPARKING PLUG

CDI UNIT

IGNITION SWITCH

HORN SWITCH

INDICATOR SWITCH

DIP SWITCH

RH FRONT INDICATOR

SPEEDO METER LAMPS

NEUTRAL INDICATOR LAMP

HOT OR WARNING LAMP

HIGH BEAM INDICATOR

HEADLAMP

LH FRONT INDICATOR

XL250 S ONLY

Wiring diagram – US XL250/500 S 1978 to 1980 models

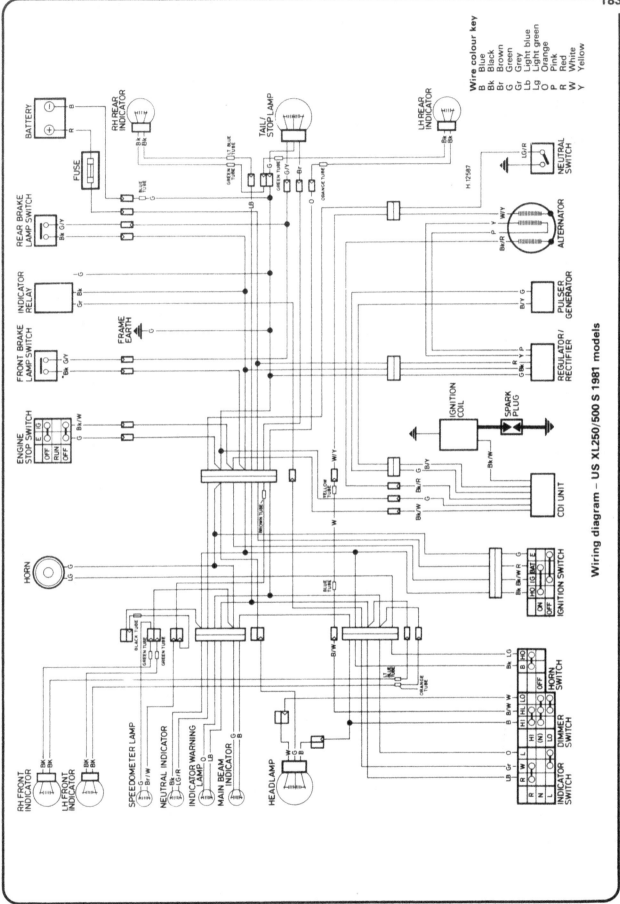

Wiring diagram – US XL250/500 S 1981 models

Wire colour key

Bl	Black
Br	Brown
Bu	Blue
G	Green
Gr	Grey
Lb	Light blue
Lg	Light green
O	Orange
P	Pink
R	Red
W	White
Y	Yellow

Wiring diagram – US XL250/500 R models

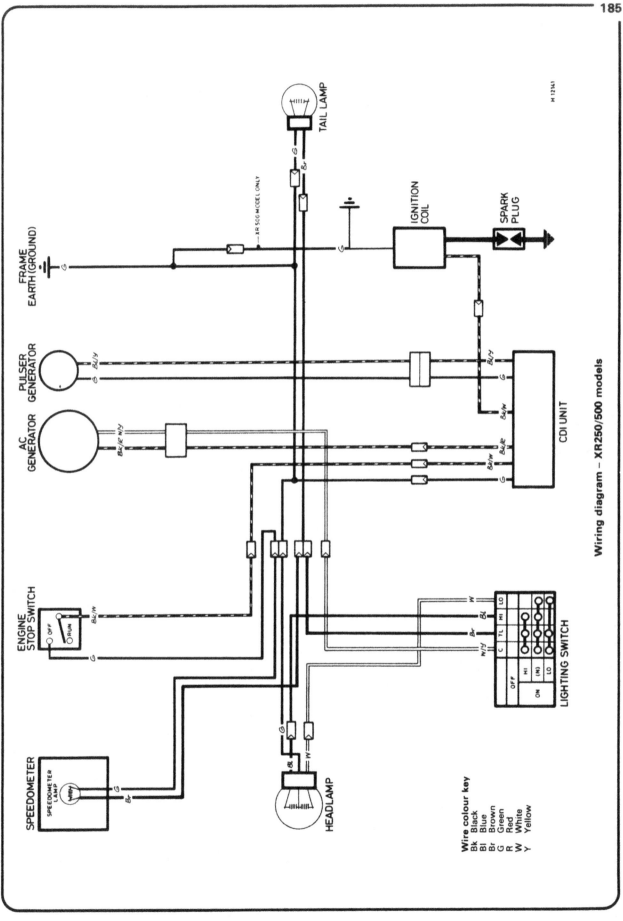

Wiring diagram – XR250/500 models

Wire colour key
Bk Black
Bl Blue
Br Brown
G Green
R Red
W White
Y Yellow

TAIL
LAMP

G
Br

H.12586

B
BK/R

ALTERNATOR

Wire colour key
B   Blue
Bk   Black
Br   Brown
G   Green
R   Red
W   White
Y   Yellow

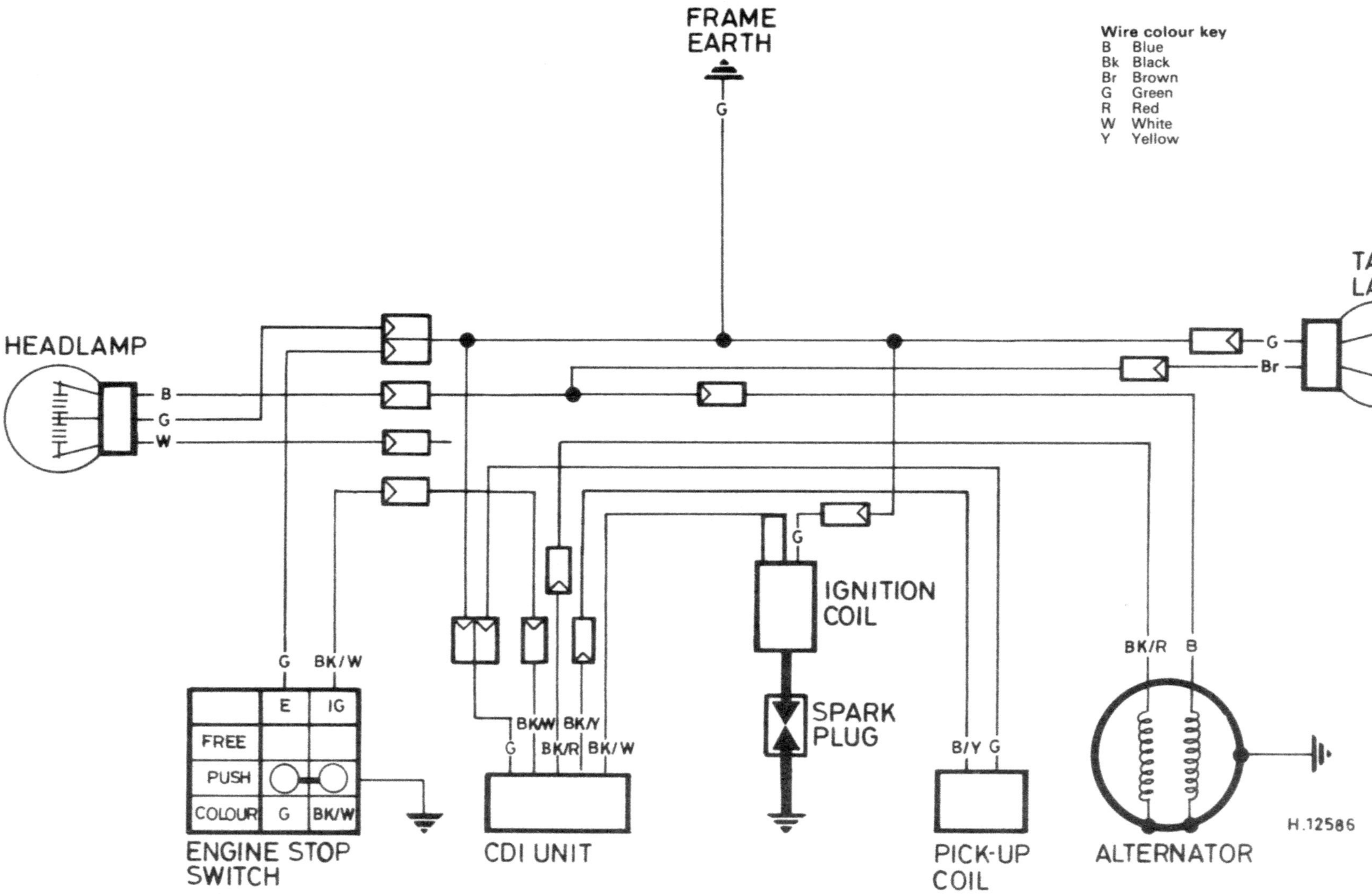

PICK-UP
COIL

G
B/Y

IGNITION
COIL

SPARK
PLUG

G

FRAME
EARTH

G

BK/W BK/Y
BK/R BK/W

G

CDI UNIT

HEADLAMP

B
G
W

BK/W

G

IG

E

FREE   PUSH   COLOUR

BK/W

G

ENGINE STOP
SWITCH

Wiring diagram – XR250/500 R models

Conversion factors

Length (distance)

Inches (in)	X	25.4	=	Millimetres (mm)	X	0.0394	= Inches (in)
Feet (ft)	X	0.305	=	Metres (m)	X	3.281	= Feet (ft)
Miles	X	1.609	=	Kilometres (km)	X	0.621	= Miles

Volume (capacity)

Cubic inches (cu in; in³)	X	16.387	=	Cubic centimetres (cc; cm³)	X	0.061	= Cubic inches (cu in; in³)
Imperial pints (Imp pt)	X	0.568	=	Litres (l)	X	1.76	= Imperial pints (Imp pt)
Imperial quarts (Imp qt)	X	1.137	=	Litres (l)	X	0.88	= Imperial quarts (Imp qt)
Imperial quarts (Imp qt)	X	1.201	=	US quarts (US qt)	X	0.833	= Imperial quarts (Imp qt)
US quarts (US qt)	X	0.946	=	Litres (l)	X	1.057	= US quarts (US qt)
Imperial gallons (Imp gal)	X	4.546	=	Litres (l)	X	0.22	= Imperial gallons (Imp gal)
Imperial gallons (Imp gal)	X	1.201	=	US gallons (US gal)	X	0.833	= Imperial gallons (Imp gal)
US gallons (US gal)	X	3.785	=	Litres (l)	X	0.264	= US gallons (US gal)

Mass (weight)

Ounces (oz)	X	28.35	=	Grams (g)	X	0.035	= Ounces (oz)
Pounds (lb)	X	0.454	=	Kilograms (kg)	X	2.205	= Pounds (lb)

Force

Ounces-force (ozf; oz)	X	0.278	=	Newtons (N)	X	3.6	= Ounces-force (ozf; oz)
Pounds-force (lbf; lb)	X	4.448	=	Newtons (N)	X	0.225	= Pounds-force (lbf; lb)
Newtons (N)	X	0.1	=	Kilograms-force (kgf; kg)	X	9.81	= Newtons (N)

Pressure

Pounds-force per square inch (psi; lbf/in²; lb/in²)	X	0.070	=	Kilograms-force per square centimetre (kgf/cm²; kg/cm²)	X	14.223	= Pounds-force per square inch (psi; lbf/in²; lb/in²)
Pounds-force per square inch (psi; lbf/in²; lb/in²)	X	0.068	=	Atmospheres (atm)	X	14.696	= Pounds-force per square inch (psi; lbf/in²; lb/in²)
Pounds-force per square inch (psi; lbf/in²; lb/in²)	X	0.069	=	Bars	X	14.5	= Pounds-force per square inch (psi; lbf/in²; lb/in²)
Pounds-force per square inch (psi; lbf/in²; lb/in²)	X	6.895	=	Kilopascals (kPa)	X	0.145	= Pounds-force per square inch (psi; lbf/in²; lb/in²)
Kilopascals (kPa)	X	0.01	=	Kilograms-force per square centimetre (kgf/cm²; kg/cm²)	X	98.1	= Kilopascals (kPa)
Millibar (mbar)	X	100	=	Pascals (Pa)	X	0.01	= Millibar (mbar)
Millibar (mbar)	X	0.0145	=	Pounds-force per square inch (psi; lbf/in²; lb/in²)	X	68.947	= Millibar (mbar)
Millibar (mbar)	X	0.75	=	Millimetres of mercury (mmHg)	X	1.333	= Millibar (mbar)
Millibar (mbar)	X	0.401	=	Inches of water (inH₂O)	X	2.491	= Millibar (mbar)
Millimetres of mercury (mmHg)	X	0.535	=	Inches of water (inH₂O)	X	1.868	= Millimetres of mercury (mmHg)
Inches of water (inH₂O)	X	0.036	=	Pounds-force per square inch (psi; lbf/in²; lb/in²)	X	27.68	= Inches of water (inH₂O)

Torque (moment of force)

Pounds-force inches (lbf in; lb in)	X	1.152	=	Kilograms-force centimetre (kgf cm; kg cm)	X	0.868	= Pounds-force inches (lbf in; lb in)
Pounds-force inches (lbf in; lb in)	X	0.113	=	Newton metres (Nm)	X	8.85	= Pounds-force inches (lbf in; lb in)
Pounds-force inches (lbf in; lb in)	X	0.083	=	Pounds-force feet (lbf ft; lb ft)	X	12	= Pounds-force inches (lbf in; lb in)
Pounds-force feet (lbf ft; lb ft)	X	0.138	=	Kilograms-force metres (kgf m; kg m)	X	7.233	= Pounds-force feet (lbf ft; lb ft)
Pounds-force feet (lbf ft; lb ft)	X	1.356	=	Newton metres (Nm)	X	0.738	= Pounds-force feet (lbf ft; lb ft)
Newton metres (Nm)	X	0.102	=	Kilograms-force metres (kgf m; kg m)	X	9.804	= Newton metres (Nm)

Power

Horsepower (hp)	X	745.7	=	Watts (W)	X	0.0013	= Horsepower (hp)

Velocity (speed)

Miles per hour (miles/hr; mph)	X	1.609	=	Kilometres per hour (km/hr; kph)	X	0.621	= Miles per hour (miles/hr; mph)

Fuel consumption*

Miles per gallon, Imperial (mpg)	X	0.354	=	Kilometres per litre (km/l)	X	2.825	= Miles per gallon, Imperial (mpg)
Miles per gallon, US (mpg)	X	0.425	=	Kilometres per litre (km/l)	X	2.352	= Miles per gallon, US (mpg)

Temperature

Degrees Fahrenheit = (°C x 1.8) + 32

Degrees Celsius (Degrees Centigrade; °C) = (°F - 32) x 0.56

*It is common practice to convert from miles per gallon (mpg) to litres/100 kilometres (l/100km), where mpg (Imperial) x l/100 km = 282 and mpg (US) x l/100 km = 235

English/American terminology

Because this book has been written in England, British English component names, phrases and spellings have been used throughout. American English usage is quite often different and whereas normally no confusion should occur, a list of equivalent terminology is given below.

English	American	English	American
Air filter	Air cleaner	Number plate	License plate
Alignment (headlamp)	Aim	Output or layshaft	Countershaft
Allen screw/key	Socket screw/wrench	Panniers	Side cases
Anticlockwise	Counterclockwise	Paraffin	Kerosene
Bottom/top gear	Low/high gear	Petrol	Gasoline
Bottom/top yoke	Bottom/top triple clamp	Petrol/fuel tank	Gas tank
Bush	Bushing	Pinking	Pinging
Carburettor	Carburetor	Rear suspension unit	Rear shock absorber
Catch	Latch	Rocker cover	Valve cover
Circlip	Snap ring	Selector	Shifter
Clutch drum	Clutch housing	Self-locking pliers	Vise-grips
Dip switch	Dimmer switch	Side or parking lamp	Parking or auxiliary light
Disulphide	Disulfide	Side or prop stand	Kick stand
Dynamo	DC generator	Silencer	Muffler
Earth	Ground	Spanner	Wrench
End float	End play	Split pin	Cotter pin
Engineer's blue	Machinist's dye	Stanchion	Tube
Exhaust pipe	Header	Sulphuric	Sulfuric
Fault diagnosis	Trouble shooting	Sump	Oil pan
Float chamber	Float bowl	Swinging arm	Swingarm
Footrest	Footpeg	Tab washer	Lock washer
Fuel/petrol tap	Petcock	Top box	Trunk
Gaiter	Boot	Torch	Flashlight
Gearbox	Transmission	Two/four stroke	Two/four cycle
Gearchange	Shift	Tyre	Tire
Gudgeon pin	Wrist/piston pin	Valve collar	Valve retainer
Indicator	Turn signal	Valve collets	Valve cotters
Inlet	Intake	Vice	Vise
Input shaft or mainshaft	Mainshaft	Wheel spindle	Axle
Kickstart	Kickstarter	White spirit	Stoddard solvent
Lower leg	Slider	Windscreen	Windshield
Mudguard	Fender		

Index